Topology Control in Wireless Ad Hoc and Sensor Networks

Topology Control in Wireless Ad Hoc and Sensor Networks

Paolo Santi

Istituto di Informatica e Telematica del CNR – Italy

John Wiley & Sons, Ltd

Copyright © 2005   John Wiley & Sons Ltd, The Atrium, Southern Gate, Chichester,
West Sussex PO19 8SQ, England

Telephone (+44) 1243 779777

Email (for orders and customer service enquiries): cs-books@wiley.co.uk
Visit our Home Page on www.wiley.com

This publication is designed to provide accurate and authoritative information in regard to the subject matter covered. It is sold on the understanding that the Publisher is not engaged in rendering professional services. If professional advice or other expert assistance is required, the services of a competent professional should be sought.

Other Wiley Editorial Offices

John Wiley & Sons Inc., 111 River Street, Hoboken, NJ 07030, USA

Jossey-Bass, 989 Market Street, San Francisco, CA 94103-1741, USA

Wiley-VCH Verlag GmbH, Boschstr. 12, D-69469 Weinheim, Germany

John Wiley & Sons Australia Ltd, 42 McDougall Street, Milton, Queensland 4064, Australia

John Wiley & Sons (Asia) Pte Ltd, 2 Clementi Loop #02-01, Jin Xing Distripark, Singapore 129809

John Wiley & Sons Canada Ltd, 22 Worcester Road, Etobicoke, Ontario, Canada M9W 1L1

Wiley also publishes its books in a variety of electronic formats. Some content that appears in print may not be available in electronic books.

Library of Congress Cataloging-in-Publication Data

Santi, Paolo.
 Topology control in wireless ad hoc and sensor networks / Paolo Santi.
 p. cm.
 Includes bibliographical references and index.
 ISBN-13: 978-0-470-09453-2 (cloth : alk. paper)
 ISBN-10: 0-470-09453-2 (cloth : alk. paper)
 1. Wireless communication systems. 2. Wireless LANs. 3. Sensor
networks. I. Title.
TK5103.2.S258 2006
004.6′8–dc22
                                   2005013736

British Library Cataloguing in Publication Data

A catalogue record for this book is available from the British Library

ISBN-13 978-0-470-09453-2 (HB)
ISBN-10 0-470-09453-2 (HB)

Typeset in 10/12pt Times by Laserwords Private Limited, Chennai, India

This book is printed on acid-free paper responsibly manufactured from sustainable forestry in which at least two trees are planted for each one used for paper production.

To my wife Elena,
my daughter Bianca,
and my children to be

To my families

Contents

About the Author

Paolo Santi is Researcher at the Istituto di Informatica e Telematica del CNR in Pisa, Italy, a position he has held since 2001. He received the 'Laurea' Degree and the PhD in Computer Science from the University of Pisa in 1994 and 2000 respectively. During his career, he visited the School of Electrical and Computer Engineering, Georgia Institute of Technology, in 2001, and the Department of Computer Science, Carnegie Mellon University, in 2003.

During his PhD studies, Dr. Santi's research activity focused on fault-tolerant computing in multiprocessor systems. Starting from 2001, his research interests shifted to wireless ad hoc networking, with particular focus on the investigation of fundamental network properties such as connectivity, network lifetime, and mobility modeling, and on the design of energy-efficient protocols.

Dr. Santi has contributed more than twenty papers in the field of wireless ad hoc and sensor networking, and has been involved in the organizational and technical committee of several conferences in the field. Dr. Santi is a member of ACM and SIGMOBILE.

Preface

The idea of this book was conceived in September 2003, in San Diego, CA, when I presented a tutorial on topology control at the ACM Mobicom conference. After the tutorial, Birgit Gruber approached me and enthusiastically suggested to me the idea of writing a book on topology control. She needed little effort to convince me indeed, since I found the idea very appealing.

The material and organization of this book have been adapted from the tutorial I presented at ACM Mobicom 2003, and later on at ACM MobiHoc 2004. In turn, the tutorial finds its origin in a survey paper on topology control that I wrote at the beginning of 2003, which is still in technical report form (the processing time of some journals is actually longer than the time needed to write a book...).

The aim of this book is to provide a unique reference resource on topology control in wireless ad hoc and sensor networks, a topic that has been a subject of intensive research in recent years. Indeed, this research field is far from being settled, and several new results and proposals are being published. This explains why writing a book on topology control has been very challenging for me. I have done my best to include in the book the most significant results and findings in the field, while at the same time describing in detail the many problems that are still to be solved. While I have tried to be as exhaustive as I could in presenting the topology control approaches introduced in the literature, the reader should bear in mind that what is reported in this book is a picture of this research field taken at the beginning of year 2005.

Audience

This book is intended for graduate students, researchers, and practitioners who are interested in acquiring a global view of the set of techniques and protocols that have been referred to as 'topology control' in the literature. More in general, the book can serve as a reference resource for researchers, engineers, and developers working in the field of wireless ad hoc and sensor networking.

While I have tried to make the book as self-contained as possible, some rudimentary knowledge of concepts of networking protocols, distributed systems, computational complexity, graph theory, and probability theory is required.

Book Overview

The material contained in this book is organized as follows.

The first part of the book (Introduction) presents introductory material that is preparatory for what is described in the rest of the book.

Chapter 1 gives a short introduction to wireless ad hoc and sensor networks, describing some of the possible applications that these technologies will make available in a near future. The chapter also discusses the many technical challenges that are still to be solved before a large-scale deployment of wireless multihop networks can actually take place.

Chapter 2 introduces the wireless network model that will be used in the rest of the book. To model a complex system like a wireless multihop network, we need several submodels: a model for a single wireless channel (Section 2.1), a model for describing all the wireless channels in the network (Section 2.2), a model for the node energy consumption (Section 2.3), and a model for node mobility (Section 2.4).

Chapter 3 tries to explain what motivated researchers to study topology control techniques. In particular, it presents simple examples showing the potential of topology control in reducing node energy consumption (Section 3.1.1) and in increasing the network traffic–carrying capacity (Section 3.1.2). The chapter also provides a first informal definition of *topology control* (TC), clarifying my personal interpretation (and the one that will be used in this book) of what is topology control, and what is *not* topology control (e.g. power control and clustering techniques) (Section 3.2). After having discussed a possible taxonomy of the many approaches to the TC problem proposed in the literature (Section 3.3), the chapter ends with a discussion on how TC mechanisms can be integrated into the network protocol stack (Section 3.4). Chapter 3 concludes the first part of the book, Introduction.

The second part of the book, The Critical Transmitting Range, treats the simplest possible form of topology control: all the nodes are assumed to have same transmitting range r, and the problem is how to choose r in such a way that certain network properties are satisfied.

Chapter 4 considers the case in which the network nodes are stationary, and the target network property is connectivity. After having formally characterized which is the critical value of r in this setting, we consider networks with dense (Section 4.1) and sparse (Section 4.2) node deployment. Then, we consider the case of nonrectangular shapes of the deployment region and/or of nonuniform node distribution (Section 4.3). The chapter ends with a discussion on what changes in the picture if the radio coverage area is not a perfect circle (Section 4.4).

Chapter 5 considers the case of mobile networks, and it discusses the implications of node mobility on the characterization of the critical range for connectivity.

Finally, Chapter 6, which ends Part II of this book, considers the different target network properties for which the critical range value is investigated, such as k-connectivity (Section 6.1), connectivity with Bernoulli nodes (Section 6.2), and sensing coverage (Section 6.3).

The third part of the book, Topology Optimization Problems, addresses several topology optimization problems. In these problems, it is typically assumed that node positions are known to a centralized observer. Given this information, the observer has the goal of identifying a certain 'optimal' topology, where the definition of 'optimal' depends on the target property considered.

The first problem considered is the so-called Range Assignment (RA) problem (Chapter 7): nodes can choose different transmitting ranges; the goal is to choose the ranges in such a way that the network is connected, and the energy-cost of the topology is minimized. This problem is studied first in one-dimensional networks (Section 7.2) and then in the more complex case of two- and three-dimensional networks (Section 7.3). Then, two symmetric variants of the Range Assignment problem are considered (Section 7.4). The chapter ends with a discussion of the energy efficiency of the optimal topologies for the various versions of the RA problem (Section 7.5).

Chapter 8, which concludes Part III of this book, addresses the problem of designing energy-optimal topologies for a certain communication pattern. The communication patterns considered are point-to-point communication, a.k.a. unicast, (Section 8.1) and one-to-all communication, a.k.a. broadcast (Section 8.2).

In the fourth part of the book, Distributed Topology Control, we consider distributed approaches to the topology control problem: the goal here is to devise fully distributed protocols that build and maintain a 'reasonably good' network topology.

Chapter 9 discusses the ideal features of a distributed TC protocol (Section 9.1), high-lighting the trade-off between the quality of information available to the nodes and the quality of the topology produced by the protocol (Section 9.2). Then, it discusses the important distinction between logical and physical degree of a node in the network topology (Section 9.3).

The following chapters present some of the most relevant distributed topology control protocols introduced in the literature, grouping them on the basis of the type of information that is available to the network nodes.

Chapter 10 presents two protocols based on the assumption that nodes know their exact location and the location of the neighbors. The protocols presented are the R&M protocol (Section 10.1) and the LMST protocol (Section 10.2).

Chapter 11 presents protocols based on directional information. In particular, it introduces the CBTC protocol (Section 11.1) and the DISTRNG protocol (Section 11.2).

Chapter 12 is concerned with approaches in which nodes are assumed to know only the ID of their neighbors, and are able to order them according to some criteria (e.g. distance, or link quality). After having discussed this TC problem from a theoretical viewpoint (Section 12.1), the chapter introduces two neighbor-based topology control protocols: the KNEIGH protocol (Section 12.2) and the XTC protocol (Section 12.3).

The last chapter of Part IV of this book, Chapter 13, discusses the effect of mobility on distributed topology control protocols, revisiting the ideal features of a distributed TC protocol (Section 13.1), and providing an example showing how different TC solutions adapt to the case of mobile networks (Section 13.2). Then, it discusses the effect of node mobility on the critical number of neighbors needed to maintain the network connected (Section 13.3). The chapter ends describing how some of the existing topology control protocols deal with node mobility (Section 13.4).

Part V of the book, Toward an Implementation of Topology Control, deals with more practical issues, describing the existing TC approaches that are closer to on-the-field implementation and the several problems that are still open in the field of topology control.

Chapter 14 describes distributed TC protocols that explicitly use a typical feature of current wireless transceivers, that is, the availability of only a limited number of possible transmit power levels. The protocols presented in the chapter are the COMPOW protocol (Section 14.2), the CLUSTERPOW protocol (Section 14.3), and the KNEIGHLEV protocol (Section 14.4).

Chapter 15, which ends Part V of the book, discusses the main open research and technological problems in the field of topology control. In particular, it outlines the need for a topology control design focused on reducing radio interference between nodes (Section 15.1), and for more realistic network models (Section 15.2). Also, much research is still to be done to address the topology control problem in mobile networks (Section 15.3) and to account for the effects of multihop data traffic (Section 15.4). The chapter ends with a discussion of practical issues that must be dealt with when implementing TC mechanisms (Section 15.5).

The final part of the book, Case Study and Appendices, provides a detailed description of a case study and two Appendices.

Chapter 16 considers the problem of implementing a routing protocol in a competitive environment, in which voluntary, unselfish participation of the network nodes to the packet forwarding task cannot be taken for granted. After having described the problem (Section 16.1) and a reference application scenario (Section 16.2), the chapter presents solutions to the cooperative routing problem that do not integrate TC mechanisms (Section 16.5), and that integrate TC and routing (Section 16.6).

Finally, Appendix A introduces basic concepts and definitions of graph theory, and Appendix B introduces basic probability notions. Appendix B also provides a short overview of three applied probability theories that have been used in the analysis of the various topology control problems presented in the book: the geometric random graph theory (Section B.2), the occupancy theory (Section B.3), and the theory of continuum percolation (Section B.4).

How to Use This Book

The book is organized into six parts. Informally speaking, the first part of the book provides basic concepts and definitions related to topology control that will be used in the rest of the book. While a reader who is familiar with the field of wireless ad hoc and sensor networks can probably skip Chapter 1, he (or she) should probably not miss Chapter 2, which introduces the network model used in the book.

After the introductory material, the topology control problem is approached firstly from a theoretical viewpoint (Part II and Part III), and then from a more practical viewpoint (Part IV and V).

The last part of the book contains an interesting case study and two appendices. The appendices are intended to provide a unique reference point for the concepts of graph theory (Appendix A) and elementary and applied probability (Appendix B) used in the book: if the reader is not sure about a certain graph theory or probability theory notion mentioned somewhere in the text, he (or she) can refer to the appropriate appendix and get it clarified. With a similar purpose, I have included an exhaustive list of the many acronyms and abbreviations used in the book.

Although, in general, topology control techniques can be used both in ad hoc and in sensor networks, some of them are more useful for application in sensor networks (Chapters 4, 6, 7, 8, 10), and others for application in ad hoc networks (Chapters 5, 11, 12, 13, 14, 16).

A reader with a background in computer science will probably be more comfortable with Part II, Part III, and Part IV of this book, while a reader with a background in engineering will probably be more comfortable with Part IV and Part V of the book. A reader with a background in applied mathematics will probably be interested in Part II and Part III of this book and Section 12.1.

Acknowledgments

There are several persons without whose support and contribution this book would have not been possible.

A first thought is for Birgit Gruber of Wiley, who contacted me in San Diego when I was presenting a tutorial on topology control, and suggested to me the idea of writing a book on this topic. Her enthusiasm was fundamental to convince me of the idea, which resulted a year and half later in this book. I also wish to thank all the staff at Wiley (Joanna Tootill and Julie Ward – I hope not to have forgotten anybody) for their assistance during the writing and the production phase of the book.

I am deeply grateful to the colleagues who shared with me the exciting task of studying the realm of topology control in these years: Doug Blough, Giovanni Resta, Mauro Leoncini, Christian Bettstetter and Stephan Eidenbenz. Much of the material presented in this book is the fruit of our collaboration. Doug also first suggested to me the idea of writing a survey paper on topology control, which, as I have explained above, can be considered as the very origin of this book. Giovanni also provided me Figure 9.1 and Figure 15.2. Christian also read a draft version of Chapters 5, and gave me many useful suggestions to improve it. To all of them I am indebted.

Pisa,
May 2005

Paolo Santi

List of Abbreviations

A.A.S.	Asymptotically Almost Surely
ACK	Acknowledgment
AoA	Angle of Arrival
AODV	Ad hoc On-demand Distance Vector
BIP	Broadcast Incremental Power
CBTC	Cone-Based Topology Control
CCR	Critical Coverage Range
CDMA	Code Division Multiple Access
CLUSTERPOW	CLUSTERed POWer
CNN	Critical Neighbor Number
COMPOW	COMmon POWer
CSMA-CA	Carrier Sense Multiple Access–Collision Avoidance
CTR	Critical Transmitting Range
CTS	Clear To Send
DistRNG	Distributed Relative Neighborhood Graph
DSDV	Dynamic destination Sequenced Distance Vector
DSR	Dynamic Source Routing
DT	Delaunay Triangulation
EMST	Euclidean Minimum Spanning Tree
FLSS	Fault-tolerant Local Spanning Subgraph
GG	Gabriel Graph
GPS	Global Positioning System
GRG	Geometric Random Graph
KNeigh	K Neighbors
KNeighLev	K Neighbors Level-based
ISN	Increase Symmetric Neighbors
LAN	Local Area Network
LILT	Local Information Link-state Topology
LINT	Local Information No Topology
LMST	Local Minimum Spanning Tree
LOS	Line Of Sight
MAC	Medium Access Control
MST	Minimum Spanning Tree
NAP	Neighbor Addition Protocol
NAV	Network Allocation Vector
NDP	Neighbor Discovery Protocol

NRP	Neighbor Reduction Protocol
PDA	Personal Digital Assistant
PDF	Probability Density Function
PSTN	Public Switched Telephone Network
QoS	Quality of Service
RA	Range Assignment
RF	Radio Frequency
R&M	Rodoplu and Meng
RNG	Relative Neighborhood Graph
RSSI	Received Signal Strength Indicator
RTS	Request To Send
RWP	Random WayPoint
SINR	Signal to Noise Ratio
TC	Topology Control
ToA	Time of Arrival
VCG	Vickrey Clarke Groves
wCNN	weak Critical Neighbor Number
W.H.P.	With High Probability
XTC	eXtreme Topology Control
YG	Yao Graph

List of Figures

List of Tables

Part I

Introduction

1

Ad Hoc and Sensor Networks

1.1 The Future of Wireless Communication

Recent emergence of affordable, portable wireless communication and computation devices and concomitant advances in the communication infrastructure have resulted in the rapid growth of mobile wireless networks. On one hand, this has led to the exponential growth of the cellular network, which is based on the combination of wired and wireless technologies. Nowadays, the number of cellular network users is approaching two billion worldwide (expected at end 2005). Although the research and development efforts devoted to traditional wireless networks are still considerable, the interest of the scientific and industrial community in the realm of telecommunications has recently shifted to more challenging scenarios in which a group of mobile units equipped with radio transceivers communicate without any fixed infrastructure.

1.1.1 Ad hoc networks

Ad hoc networks are the ultimate frontier in wireless communication. This technology allows network nodes to communicate directly to each other using wireless transceivers (possibly along multihop paths) without the need for a fixed infrastructure. This is a very distinguishing feature of ad hoc networks with respect to more traditional wireless networks, such as cellular networks and wireless LAN, in which nodes (for instance, mobile telephone users) communicate with each other through base stations (wired radio antennae).

Ad hoc networks are expected to revolutionize wireless communications in the next few years: by complementing more traditional network paradigms (Internet, cellular networks, satellite communications), they can be considered as the technological counterpart of the concept of ubiquitous computing. By exploiting ad hoc wireless technology, various portable devices (cellular phones, PDAs, laptops, pagers, and so on) and fixed equipment (base stations, wireless Internet access points, etc.) can be connected together, forming a sort of 'global', or 'ubiquitous', network.

Application scenarios in which the adoption of ad hoc networking technologies might prove useful abound. For instance, consider the following situation. A terrible earthquake has

Topology Control in Wireless Ad Hoc and Sensor Networks   P. Santi
© 2005 John Wiley & Sons, Ltd

devastated the city of Futuria destroying, among other things, most of the communication infrastructure (wired phone lines, base stations for cellular networks, and so on). Several rescue teams (firefighters, police, medical teams, volunteers, and so on) are working on the disaster scene to save people from wreckage and to assist the injured. To provide a better assistance to the population, the efforts of the rescue teams should be coordinated. Clearly, a coordinate action can be achieved only if rescuers are able to communicate, both within their team (e.g. a policeman with other policemen) and with members of the other teams (e.g. a firefighter calling a doctor for assistance). With currently available technology, coordinating rescuers' efforts when the fixed communication infrastructure is severely damaged is very difficult: even if team members are equipped with walkie-talkies or similar devices, when no access to the fixed infrastructure is available, only communication between nearby rescuers is possible. Thus, one of the priorities in present-day disaster management is to reinstall the communication infrastructure as quickly as possible, which is typically done by repairing the damaged structures and by deploying temporary communication equipment (e.g. vans equipped with a radio antenna).

The situation would be considerably different if technologies based on ad hoc networking were available: by using fully decentralized, multihop wireless communication, even relatively distant rescuers would be able to communicate, provided there exist other team members in between them acting as communication relay. Since a disaster area is typically quite densely populated with rescuers, citywide (or even metropolitanwide) communication would be possible, allowing a successful coordination of the rescue efforts without the need for reestablishing the fixed communication infrastructure.

The above-described example outlines the features of a typical ad hoc network application scenario:

- *Heterogeneous network*: A typical ad hoc network is composed of heterogeneous devices. For instance, in the scenario described above, in general the various teams working on the disaster area are equipped with different types of devices: cell phones, PDAs, walkie-talkies, laptops, and so on. For a successful setup of the communication network, it is fundamental that these diverse types of devices be able to communicate with each other.

- *Mobility*: In a typical ad hoc network, most of the nodes are mobile. This is the case, for instance, of the rescuers working in a disaster scenario as described above.

- *Relatively dispersed network*: The adoption of the ad hoc networking paradigm is justified when the nodes composing the network are geographically dispersed. In fact, if network nodes are very close to each other, 1-hop wireless communication is usually possible and no multihop communication between nodes is necessary.

Potential application of wireless ad hoc networks are numerous. Among them, we cite the following:

- *Fast traffic info. delivery on highways and urban areas*: Highways and urban areas can be equipped with fixed radio transmitters, which broadcast traffic information to cars equipped with GPS receivers passing close to a transmitter. In turn, the cars themselves act as relay of information so that the traffic updates can quickly reach faraway drivers. As compared to traditional radio traffic info delivery, this technology will provide a much more accurate (localized) and faster service.

– *Ubiquitous Internet access*: In a very near future (in part, this is already a reality), public areas such as airports, stations, shopping malls, and so on, will be equipped with wireless Internet access points. By using the portable devices of other users as wireless bridges, Internet access can be extended to virtually the entire urban area.

– *Delivery of location-aware information*: By using fixed radio transmitters (for instance, the same transmitters used to broadcast traffic updates), location-aware information can be delivered to the interested users. Examples of location-aware information are tourist information, shows and events in the surrounding, information on shops/restaurants in the area, and so on.

1.1.2   Wireless sensor networks

Wireless sensor networks (WSNs for short) are a particular type of ad hoc network, in which the nodes are 'smart sensors', that is, small devices (approximately the size of a coin) equipped with advanced sensing functionalities (thermal, pressure, acoustic, and so on, are examples of such sensing abilities), a small processor, and a short-range wireless transceiver. In this type of network, the sensors exchange information on the environment in order to build a global view of the monitored region, which is made accessible to the external user through one or more gateway node(s).

Sensor networks are expected to bring a breakthrough in the way natural phenomena are observed: the accuracy of the observation will be considerably improved, leading to a better understanding and forecasting of such phenomena. The expected benefits to the community will be considerable.

As in the case of ad hoc networks, to give a better idea of the potential of WSN technology, we describe in detail a sample application scenario. Consider a situation in which a WSN is used to monitor a vast and remote geographical region, in such a way abnormal events (e.g. a forest fire) can be quickly detected. In this scenario, smart sensors, each equipped with a battery, and significant processing and wireless communication capabilities, are placed in strategic positions, for example, on the top of a hill or in locations with wide view. Each sensor covers a few hectares area and can communicate with sensors in the surrounding. The sensor node gathers atmospheric data (temperature, pressure, humidity, wind velocity and direction) and analyzes atmosphere makeup to detect particular particles (e.g. ash). Furthermore, each sensor node is equipped with an infrared camera, which is able to detect thermal variations. Every sensor knows its geographic position, expressed in terms of degree of latitude and longitude. This can be accomplished either by equipping every node with a GPS receiver, or, since in this scenario sensor position is fixed, by setting the position in a sensor register at the time of deployment. Periodically, sensors exchange data with neighboring nodes in order to detect unusual situations that could be caused, for instance, by a starting fire (e.g. temperature at a sensor much higher than those of the neighbors). These 'routine' data are aggregated and propagated throughout the network and can be gathered by the external operator to collect atmospheric data (e.g. to check the air quality). When a potentially dangerous situation is detected (for instance, the infrared camera detects a rapid thermal increase in a certain zone), an emergency procedure is started: the sensor node that has detected the abnormal condition communicates with its neighbors in order to verify whether the same condition has been detected by other sensors; then, it tries to accurately determine the geographic position of the hazard (if the same abnormal situation has been detected by other sensors, this can be accomplished using triangulation techniques; furthermore, the information on the wind velocity and direction can be useful both in the localization of the fire and in forecasting the direction of its propagation); once the position of the fire has been determined, an alarm message containing the fire's geographic coordinates and (possibly) its propagation direction is disseminated with the maximum priority. This way, the external operator (for instance, a park ranger equipped with a portable device) is promptly alerted of the presence of fire, of its position, and of the forecasted propagation direction of the fire, and can intervene quickly.

The fire-detection application scenario is summarized in Figure 1.1. We remark that this scenario has several interesting features, such as reduced impact on the environment (since sensor nodes have wireless transceivers, no wiring is needed), accuracy of coverage, and prompt alerting of the human operator.

The above-described example outlines the features of a typical WSN application scenario:

- *Homogeneous network*: Differing from the case of ad hoc networks, a WSN is typically composed of nodes with the same features, especially for what concerns the communication apparatus. A partial exception to this rule is when different types of smart sensor nodes are used in the same network: for instance, a few 'super nodes' (with more memory and/or with a longer transmitting range) could be used in combination with standard sensor nodes to increase the network monitoring ability. However, also in this case the number of different device classes used in the network is very limited (2–3 at most).

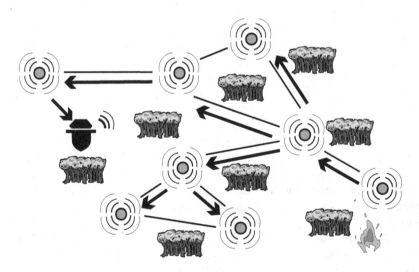

Figure 1.1 Sensor network used for prompt fire detection. When a fire is detected, an alarm message (arrow) is generated by the sensor node(s) that detected the fire. The message is then propagated in the network until it reaches a park ranger.

Table 1.1 Comparison of typical features of wireless ad hoc and sensor networks

Ad hoc Networks	WSNs
Heterogeneous devices	Homogeneous devices
Mobile nodes	Stationary nodes
Dispersed network	Dispersed network
	Large network size

- *Stationary or quasistationary network*: Differing from the case of ad hoc networks, nodes composing a WSN are typically stationary, or at most slowly moving. Given the very wide range of WSN applications, exceptions to this rule are possible. This is the case, for instance, of a sensor network used to track animal movements.

- *Relatively dispersed network*: this feature is in common with ad hoc networks: a wireless sensor network is typically formed by nodes that are dispersed in a relatively large geographical region, so that 1-hop communication between nodes is, in general, not possible.

- *Large network size*: Typically, the number of nodes composing a WSN is quite large, ranging from few tens to thousands of nodes.

The differences/similarities between ad hoc and sensor networks are summarized in Table 1.1.

Among the many possible WSN application scenarios, we cite the following:

- *Ocean temperature monitoring for improved weather forecast*: It is known that the evolution of weather conditions is strongly influenced by the temperature of large water masses such as the oceans. However, nowadays our ability to perform a large-scale monitoring of the ocean temperature is scarce. Sensor networks can be used for this purpose. By dropping a large number of tiny sensors into the sea, water temperature and ocean currents can be accurately monitored, helping the scientists in the task of providing more accurate weather forecast.

- *Intrusion detection*: Camera-equipped sensors can be used to form a network that monitors an area with restricted access. If the network is properly deployed, intruders can be detected and an alarm message quickly propagated to the external observer.

- *Avalanche prediction*: Sensors equipped with location devices (such as GPS) can be used to monitor the movements of large snow masses, thus allowing a more accurate avalanche prediction.

1.2 Challenges

Although the technology for ad hoc and sensor networks is relatively mature, the applications arc almost completely lacking. This is in part due to the fact that some of the problems related to ad hoc/sensor networking are still unsolved. In this section, we describe the state of progress of the current ad hoc and sensor network technology, and the main challenges that face the ad hoc/sensor network designer.

1.2.1   Ad hoc networks

Wireless ad hoc networks have attracted the attention of researchers in academia and industry in the last few years. As a result of this considerable research activity, the basic mechanisms that enable wireless ad hoc communication have been designed and standardized. Just to cite the most popular examples, IEEE 802.11 (IEEE 1999) and Bluetooth (Bluetooth 1999) are communication standards that are implemented in a variety of commercial wireless equipment, and that allows infrastructure-less wireless communication between mutually compliant devices. Thus, wireless, multihop communication between different types of devices such as cell phones, laptops, PDAs, smart appliances, and so on, is possible with currently available technology.

Despite the fact that the technology for ad hoc network exists, applications based on the ad hoc networking paradigm are almost completely lacking. This is because many of the challenges to be faced for a practical implementation of ad hoc network services are still to be solved. The main such challenges are the following:

- *Energy conservation*: Since units in ad hoc networks are typically battery equipped, one of the primary design goals is to use this limited amount of energy as efficiently as possible.

- *Unstructured and/or time-varying network topology*: Since the network nodes can, in principle, be arbitrarily placed in a certain region and are typically mobile, the topology of the graph that represents the wireless communication links between the nodes is usually unstructured. Furthermore, the network topology may vary with time, because of node mobility and/or failure. In these conditions, optimizing the performance of ad hoc network protocols is a very difficult task.

- *Low-quality communications*: Communication on a wireless channel is, in general, much less reliable than in a wired channel. Furthermore, the quality of communication is influenced by environmental factors (weather conditions, presence of obstacles, interference with other radio networks, etc.), which are time varying. Thus, applications for ad hoc networks should be resilient to dramatically varying link conditions, tolerating also nonnegligible off-service time intervals of the wireless link.

- *Resource-constrained computation*: Ad hoc networks are characterized by scarce resource availability; in particular, energy and network bandwidth are available in very limited amounts as compared to more traditional network paradigms. Protocols for ad hoc networks must strive to provide the desired performance level in spite of the few available resources.

- *Scalability*: In some ad hoc network scenarios, the network can be composed of hundreds or thousands of nodes. This means that protocols for ad hoc networking must be able to operate efficiently in the presence of a very large number of nodes also.

In case of ad hoc networks used for 'ubiquitous' networking, the following issues must also be addressed:

- *Interoperability*: In the 'ubiquitous' networking scenario described in Section 1.1.1, data should travel through the most diverse type of networks: ad hoc, cellular, satellite, wireless LAN, PSTN, Internet, and so on. Ideally, the user should smoothly switch from one network to the other without interrupting her applications. Implementing this sort of 'network handoff' is a very challenging task.

- *Definition of a feasible business model*: Currently, accounting in wireless networks (cellular, and commercial wireless Internet access) is done at the base station, that is, using a centralized infrastructure. Furthermore, roaming is allowed only within networks of the same type (e.g. cell phone roaming when the user is in a foreign country). In the 'ubiquitous' scenario, it is still not clear which infrastructure should perform billing and which rules should be used to regulate roaming between different types of networks.

- *Stimulate cooperation between nodes*: When designing a certain network protocol, it is usually assumed that all the nodes in the network voluntarily participate in the protocol execution. In some ad hoc network application scenarios, network nodes are owned by different authorities (private users, professionals, profit and/or nonprofit organizations, and so on), and voluntary participation in the protocol execution cannot be taken for granted. Thus, network nodes must be somehow stimulated to behave according to the protocol specifications. The issue of stimulating cooperation between nodes is treated in some detail in Chapter 16.

1.2.2 Wireless sensor networks

In a manner similar to ad hoc networks, WSNs also have attracted the attention of both the academic and the industrial research community in the last few years. Firstly, a number of smart sensor prototypes have been designed and implemented by the academic research community. The most famous of such prototypes are probably the Berkeley Motes (Polastre et al. 2004) and Smart Dust (Pister 2001). Later on, many academic interdisciplinary projects have been funded (and are currently being funded) to actually deploy and utilize sensor networks. One such example is the Great Duck Island project, in which a WSN has been deployed to monitor the habitat of the nesting petrels without any human interference with animals (Mainwaring et al. 2002).

Smart sensor nodes are also being produced and commercialized by some electronic manufacturer. We cite Crossbow, a company that produces on a large scale the Motes sensor nodes developed at UC Berkeley. Other major silicon companies such as Intel, Philips, Siemens, STMicrolectronics, and so on, are interested in the WSN technology, and are developing their own smart sensor node platform.

There is also a considerable standardization activity in the field of WSNs. The most notable effort in this direction is the IEEE 802.15.4 standard currently under development, which defines the physical and MAC layer protocols for remote monitoring and control, as well as sensor network applications. ZigBee (ZigBeeAlliance 2004) is an industry consortium (currently involving more than 100 members, representing 22 countries on four continents) with the goal of promoting the IEEE 802.15.4 standard.

Currently, we are in a phase in which the technology for implementing wireless sensor networks is relatively mature but applications based on sensor networks have not been completely defined. In particular, industries strive to find significant markets for WSN applications. The most promising ones seem to be home control, building automation, industrial automation, and automotive applications (ZigBeeAlliance 2004). Nevertheless, the market for wireless sensor hardware is expected to grow at the rate of 20% per year in the next few years, which is three times the growth rate of the wired sensor market (Frost and Sullivan 2003).

In case of sensor networks also, many challenges are still to be faced before they can be deployed on a large scale. The main challenges related to WSN implementation are the following:

- *Energy conservation*: If reducing node energy consumption is important in ad hoc networks, it becomes vital in WSNs. In fact, because of the reduced sized of the sensor nodes, the battery has low capacity, and the available energy is very limited. Despite this scarcity of energy, the network is expected to operate for a relatively long time. Given that replacing/refilling batteries is usually impossible, one of the primary design goals is to use this limited amount of energy as efficiently as possible.

- *Low-quality communications*: Sensor networks are often deployed in harsh environments, and sometimes they operate under extreme weather conditions. In these situations, the quality of the radio communication might be extremely poor, and performing the requested collective sensing task might become very difficult.

- *Operation in hostile environments*: In many scenarios, sensor networks are expected to operate under critical environmental conditions. Thus, it is essential that in these cases the physical sensor nodes are carefully designed. Furthermore, the protocols for network operation should be resilient to sensor faults, which can be considered a relatively likely event.

- *Resource-constrained computation*: If the resources in ad hoc networks are scarce, the situation is even worse in WSNs. Protocols for sensor networks must strive to provide the desired QoS in spite of the few available resources.

- *Data processing*: Given the energy constraints and the relatively poor communication quality, the data collected by the sensor node must be locally compressed, and aggregated with similar data generated by neighboring nodes. This way, relatively few resources are used to communicate the data to the external observer. Since the observer is often interested in getting data with different levels of accuracy depending, for instance, on the events currently going on in the monitored region, the data aggregation mechanism should be able to provide different levels of compression/aggregation, addressing the data accuracy/resource consumption trade-off.

- *Scalability*: WSNs are typically composed of hundreds or even several thousands of nodes. Thus, the scalability of protocols for WSNs must be explicitly considered at the design stage.

- *Lack of easy-to-commercialize applications*: Nowadays, several chip makers and electronic companies have started the commercial production of sensor nodes. However, it is much more difficult for these companies to commercialize *applications* based on sensor networks. Selling applications, instead of relatively cheap sensors, would be much more profitable for industry. Unfortunately, most sensor network application scenarios are very specific, and a company would have little or no profit in developing an application for a very specific scenario since the potential buyers would be very few.

2

Modeling Ad Hoc Networks

In this chapter, we introduce a simple but widely accepted model of ad hoc network. Since sensor networks are a subclass of ad hoc networks, this model applies to this type of networks also.

2.1 The Wireless Channel

Nodes in ad hoc and sensor networks communicate through wireless transceivers. For this reason, an important building block of any model for ad hoc networks is the wireless channel model. The model presented in this section is based on the material contained in (Rappaport 2002).

A radio channel between a transmitter unit u and a receiver unit v is established if and only if the power of the radio signal received by node v is above a certain threshold, called the *sensitivity threshold*. Formally, there exists a direct wireless link between u and v if $P_r \geq \beta$, where P_r is the power of the signal received by v, and β denotes the sensitivity threshold. The exact value of β depends on the features of the wireless transceiver and on the communication data rate: for a given radio, the higher the data rate, the higher the value of β, implying a stronger requirement on the received power. In order to simplify notation, in the following we assume that β has the conventional value of 1.

The received power P_r depends on the power P_t used by u to transmit the radio signal, and on the *path loss*, which models the radio signal degradation with distance. Denoting with $PL(u, v)$ the path loss between units u and v, we can write

$$P_r = \frac{P_t}{PL(u, v)}.$$

Thus, the occurrence of a radio channel between any two network nodes can be predicted if the path loss model is known.

Modeling path loss has historically been one of the most difficult tasks of the wireless system designer. The mechanisms that regulate radio signal propagation in the environment can be grouped into three categories: *reflection*, *diffraction* and *scattering*. Reflection occurs when the electromagnetic wave hits the surface of an object that has very large dimensions when compared to the wavelength of the propagating signal. For instance, the radio signal is reflected by the surface of the earth and by large buildings and walls. Diffraction is caused by objects with very sharp edges that lie on the radio path between the transmitter and the receiver. Scattering occurs when several small objects (as compared to the signal wavelength) are in between the transmitter and the receiver of the radio signal. Typical sources of scattering are foliage, street signs, and so on. Given these mechanisms, it is clear that radio wave propagation is an extremely complex phenomenon, which is heavily influenced by environmental factors. In the following, we shortly describe the most common path loss models introduced in the literature. For a detailed treatment of this subject, the reader is referred to (Rappaport 2002).

2.1.1 The free space propagation model

This model is used to predict radio signal propagation when the path between the transmitter and the receiver is clear and unobstructed (line-of-sight, or LOS, path). Denoting with $P_r(d)$ the power of the radio signal received by a node located at distance d from the transmitter, we have

$$P_r(d) = \frac{P_t G_t G_r \lambda^2}{(4\pi)^2 d^2 L}, \tag{2.1}$$

where G_t is the transmitter antenna gain, G_r is the receiver antenna gain, L is the system loss factor not related to propagation, and λ is the wavelength in meters. Since we are not interested in the specific characteristics of the transceiver, we can simplify equation (2.1) as follows:

$$P_r(d) = C_f \cdot \frac{P_t}{d^2}, \tag{2.2}$$

where C_f (f stands for free space) is a constant that depends on the characteristics of the transceivers.

Equation (2.2) shows that the received power falloff is proportional to the square of the distance d that separates the transmitter and the receiver. Combining equation (2.2) with the sensitivity threshold, we can state that the transmitted message can be correctly received if and only if

$$d \leq \sqrt{C_f P_t}.$$

In other words, the radio coverage area of a node transmitting at power P_t is a disk of radius $\sqrt{C_f P_t}$ centered at the node.

The free space equation is valid only for values of d that are relatively far from the transmitting antenna. For values of d within the so-called *close-in distance* d_0, the path loss can be assumed to be constant.

2.1.2 The two-ray ground model

It is seldom the case that the single direct path between the transmitter and the receiver is the only physical means of propagation of the radio signal. For this reason, the free space propagation model is often inaccurate. To improve accuracy, the two-ray ground model considers two propagation paths: the directed path and a ground reflected propagation between the transmitter and the receiver (see Figure 2.1).

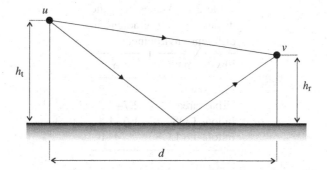

Figure 2.1 The two-ray propagation model: the radio signal sent by node u reaches node v through the direct path, and through a ground reflected path.

In the two-ray ground propagation model, the received power at distance d is given by the following formula:

$$P_r(d) = P_t G_t G_r \frac{h_t^2 h_r^2}{d^4},$$ (2.3)

where h_t is the transmitter antenna height and h_r is the receiver antenna height. If the distance between the sender and the receiver is relatively large ($d \gg \sqrt{h_t h_r}$), and abstracting the features of the radio transceivers, we can write the following simplified formula:

$$P_r(d) = C_t \cdot \frac{P_t}{d^4},$$ (2.4)

where C_t (t stands for two-ray ground) is a constant that depends on the characteristics of the radio transceivers. So, the major difference with the free space model is that the radio signal falloff in this case is proportional to the distance raised to the fourth power, instead of to the square of the distance.

Combining equation (2.4) with the sensitivity threshold, we have that the radio coverage region in the two-ray ground model is a disk of radius $\sqrt[4]{C_t P_t}$ centered at the transmitter.

2.1.3 The log-distance path model

The log-distance model has been derived combining analytical and empirical methods. Empirical methods are based on field measurements and reverse curve fitting on the experimental data.

This model, which can be seen as a generalization of both the free space and the two-ray ground model, indicates that the average long-distance path loss is proportional to the separation distance d raised to a certain exponent α, which is called the *path loss exponent*, or *distance-power gradient*. Formally,

$$P_r(d) \propto \frac{P_t}{d^\alpha}.$$ (2.5)

The radio coverage region in this model is a disk of radius proportional to $\sqrt[\alpha]{P_t}$ centered at the transmitter.

The value of α depends on the environmental conditions, and it has been experimentally evaluated in many scenarios. Table 2.1 summarizes some of these values.

Table 2.1 Values of the
distance-power gradient in
different environments

Environment	α
Free space	2
Urban area	2.7–3.5
Indoor LOS	1.6–1.8
Indoor no LOS	4–6

2.1.4 Large-scale and small-scale variations

The log-distance propagation model predicts the *average* received power at a certain distance. However, the intensity of the received signal can vary a lot from the average value. For this reason, probabilistic models have been used to account for the variability of the wireless channel. In a probabilistic propagation model, the radio coverage region is no longer a disk, since the occurrence of a wireless channel between two nodes is a random event.

Probabilistic propagation models can be divided into two classes:

– *Large-scale models*: These models predict variations of the signal intensity over large distances.

– *Small-scale models*: These models predict variations of the signal intensity over very short distance. They are also called *multipath fading* (or simply *fading*) models.

The most important large-scale model is the log-normal shadowing model, in which the path loss at distance d is modeled as a random variable with log-normal distribution (see Appendix B for a definition of log-normal distribution) centered about the mean value, which is stated in equation (2.5). The most important fading model is the Rayleigh model, which models small-scale variations of the radio signal intensity according to a random variable with Rayleigh distribution. A detailed description of probabilistic radio propagation models can be found in (Rappaport 2002).

2.2 The Communication Graph

The communication graph defines the network topology, that is, the set of wireless links that the nodes can use to communicate with each other. Given the discussion of the previous section, it is clear that the presence of a link between two units u and v in the network depends on (i) the relative distance between u and v, (ii) the transmit power used to send the data, and (iii) the surrounding environment. Since accounting for large- and small-scale variations of the radio signal is quite complicated, and renders the link model tightly coupled with a specific application scenario, in this section and in the rest of this book we will model the wireless channel using the log-distance path model, which abstracts many characteristics of the environment. This assumption is standard in research on topology control in ad hoc/sensor networks.

Let N be a set of wireless nodes, with $|N| = n$. These nodes are located in a certain bounded region R. For simplicity, we assume that R is a d-dimensional cube of side l. Formally, $R = [0, l]^d$ for some $l > 0$, where $d = 1, 2, 3$. For any $u \in N$, the *location* of u in R, denoted $L(u)$, is the position of u in R, expressed as d-dimensional coordinates. Thus, function $L : N \mapsto R$ maps every node of the network to its physical location within R. If nodes are mobile, the physical node location is time dependent. If nodes move within R – we can assume this without loss of generality – mobility can be represented by adding a further argument to L, the time instant t. Summarizing, function $L : N \times T \mapsto R$ assigns to every element of N and to any time $t \in T$ a set of d-dimensional coordinates, representing the physical node's location at time t. A d-dimensional mobile ad hoc network is then represented by the pair $M_d = (N, L)$, where N and L are defined as above.

Given a network $M_d = (N, L)$, a *range assignment* for M_d is a function RA that assigns to every element u of N a value $RA(u) \in (0, r_{\max}]$, representing its *transmitting range*. Parameter $r_{\max}$ is called the *maximum transmitting range*, and depends on the features of the radio transceivers equipping the nodes. It is usually assumed that network nodes are equipped with transceivers having similar features, that is, $r_{\max}$ is the same for all the nodes in the network. In case the network is composed of units equipped with transceivers of different type, $r_{\max}$ is intended as the maximum over the transmitting ranges of the different radios, and the definition of transmitting range is still sound.

The transmitting range of a node u denotes the range within which the data transmitted by u can be correctly received. Given the range r, the definition of the subregion of R within which correct data reception is possible depends on the network dimension: in case of one-dimensional networks, it is the segment of length $2r$ centered at u; in case of two-dimensional networks, it is the circle of radius r centered at u; in three-dimensional networks, it is the sphere of radius r centered at u (see Figure 2.2).

Note that, under the assumption that the radio signal propagates according to the log-distance path model, any transmitting range $r \in (0, r_{\max}]$ is uniquely associated with a transmit power $P_r \in (0, P_{\max}]$, where $P_{\max}$ is the maximum transmit power level of the nodes. Thus, the notions of node's transmitting range and node's transmit power are equivalent, and they will be interchangeably used in the rest of this book.

Given a network $M_d = (N, L)$ and a range assignment RA, the *communication graph* induced by RA on M_d at time t is defined as the directed graph $G_t = (N, E(t))$, where the directed edge $[u, v]$ exists if and only if $RA(u) \geq \delta(L(u, t), L(v, t))$, where $\delta(L(u, t), L(v, t))$ is the Euclidean distance between u and v at time t. In other words, the directed wireless link (u, v) exists if and only if nodes u and v are at distance of at most

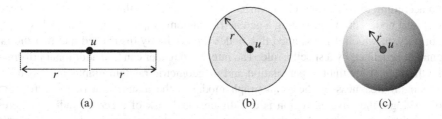

(a)   (b)   (c)

Figure 2.2   Radio coverage in one-dimensional (a), two-dimensional (b), and three-dimensional (c) networks. The covered region has radius r, and it is centered at the unit.

$RA(u)$ at time t. In this case, v is said to be a 1-hop neighbor, or *neighbor* for short, of node u. A wireless link is said to be *bidirectional*, or *symmetric*, at time t if $(u, v) \in E(t)$ and $(v, u) \in E(t)$. In this case, nodes u and v are said to be *symmetric neighbors*.

The *maxpower range assignment* is such that $RA(u) = r_{\max}$ for every node u, that is, every node in the networks transmits at maximum power. The resulting communication graph is called the *maxpower graph*, and represents the set of all possible communication links between the network nodes.

A range assignment RA is said to be *connecting at time t*, or simply *connecting*, if the resulting communication graph at time t is strongly connected, that is, if for any pair of nodes u and v, there exists at least one directed path from u to v. A range assignment in which all the nodes have the same transmitting range r, for some $0 < r \leq r_{\max}$, is called r-*homogeneous*. When the exact value of r is not relevant, the r-homogeneous range assignment is simply called *homogeneous*. Observe that the communication graph generated by a homogeneous range assignment can be considered as undirected, since $(u, v) \in E(t) \Leftrightarrow (v, u) \in E(t)$.

If the network is mobile, the range assignment may vary with time in order to preserve a certain property of the communication graph, such as connectivity. In general, we can then define a sequence of range assignments $RA_{t_1}, RA_{t_2}, \ldots$ during the network lifetime, where RA_{t_i} is the range assignment at time t_i, and the transition between range assignments is determined by an appropriate protocol.

If the network is stationary (i.e. the position of every node does not change during the entire network operational lifetime), the model described above can be simplified by making L a function of N only. Nevertheless, different range assignments can, in principle, be used during the network lifetime. The range assignment can be varied, for instance, to support different kinds of traffic (for example, in a sensor network, the type of information delivered to the external observer changes depending on the detected events), or to achieve a balanced energy consumption among network nodes. Thus, in general, the communication graph is time dependent even if the network is stationary.

The model described in this section is essentially the *point graph* model introduced in (Sen and Huson 1996). An example of two-dimensional point graph is reported in Figure 2.3. Similar graph models have been used in applied probability theories, such as *continuum percolation* and *geometric random graphs*. In the former theory, a *unit disk graph* is a graph in which every two nodes are connected with an edge if and only if they are at distance at most 1. Up to a normalization, unit disk graphs correspond to the model described in this section with homogeneous range assignment. In the theory of GRG, a set of points is distributed according to some probability distribution in a certain region. Points are then connected according to some rule (e.g. connect to all the points within distance r, or connect to the k closest nodes, etc.), generating a geometric random graph. Also, this model is a special case of ours, in which it is assumed that nodes are randomly distributed and that the range assignment is defined by a specific rule. The interested reader can find additional information on the theories of continuum percolation and of geometric random graphs in Appendix B.

The main weakness of the point graph model is the assumption of perfectly regular radio coverage: the covered region is a d-dimensional disk of a certain radius centered at the transmitter. As discussed in the previous section, this assumption is quite realistic in open air, flat environments. Unfortunately, ad hoc and sensor networks are likely to be used in very different situations, such as indoor or urban scenarios (ad hoc networks), or under

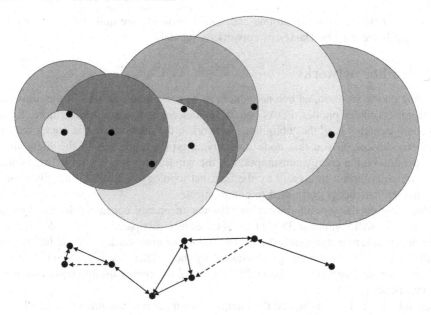

Figure 2.3 Example of two-dimensional point graph. Note that two of the links in the graph are unidirectional.

harsh conditions (sensor networks). In other words, in real-life situations, it is quite likely that the radio coverage region is highly irregular, because of the influence of walls, buildings, interference with preexisting infrastructure, and so on. However, including all these details in the network model would make it extremely complex and scenario dependent, hampering the derivation of meaningful and sufficiently general analytical results. For this reason, despite its limitations, the point graph model is widely used in the study of ad hoc network properties.

Before ending this section, we want to emphasize that the results obtained using the point graph model are useful, at least to some extent, also in situations in which the radio coverage region is known to be irregular. For instance, suppose we want to formally characterize the minimum value of r such that the r-homogeneous range assignment is connecting. If the network is two-dimensional, the value of r thus obtained can be thought of as the radius of the largest circular subarea of the actual radio coverage area. Thus, there could exist nodes that are 1-hop neighbors in reality, but that are not directly connected in the point graph model. It follows that the actual network connectivity is higher than that formally characterized using the point graph model. Clearly, the analytical characterization of r becomes less and less significant from a practical point of view as the actual radio coverage area is more and more irregular.

2.3 Modeling Energy Consumption

One of the primary concerns of the ad hoc/sensor networks designer is the efficient use of energy. Thus, it is fundamental to model the node energy consumption accurately. Since the features of typical nodes in ad hoc and sensor networks are quite different, we discuss energy models for the two classes of networks separately.

2.3.1   Ad hoc networks

Depending on the scenario, ad hoc networks can be composed of nodes of the most diverse type: laptops, cellular phones, PDAs, smart appliances, and so on. Furthermore, for many application scenarios (e.g. the 'ubiquitous network'), the network can be composed of heterogeneous devices. Given this node diversity, a typical approach in the literature is to focus attention on the energy consumption of the wireless transceiver only. This is also our choice, which is further motivated by the fact that topology control is primarily concerned with reducing the energy consumed to communicate.

Depending on the type of device, the amount of energy consumed by the transceiver varies from about 15 to about 35% of the total energy dissipated by the node. The former value refers to a laptop equipped with a IEEE 802.11 wireless card, while the latter is typical of a PDA device. Since the energy consumed by the wireless card is a significant portion of the total power dissipation in the node, optimizing the energy used to communicate is an important issue.

Several authors have measured the energy consumption of commercial 802.11 wireless cards. Typically, an IEEE 802.11 wireless card has four operational modes:

– *Idle*: The radio is turned on, but it is not used.

– *Transmit*: The radio is transmitting a data packet.

– *Receive*: The radio is receiving a data packet.

– *Sleep*: The radio is powered down.

Table 2.2 shows the power consumption of a CISCO Aironet IEEE 802.11 a/b/g card, as reported in the data sheets (Cisco 2004). Power consumption in sleep mode is not reported in the data sheets. The table also reports the nominal transmitting ranges when the card transmits at full power. As seen from the table, the nominal range depends on environmental factors (indoor or outdoor conditions) and on the data rate used to send the message.

We remark that the data reported in Table 2.2 are nominal, and can be considerably different from the actual power consumption of the wireless card. For instance, if we consider the specifics of the CISCO Aironet 350 card as reported in the data sheets, the *sleep* : *idle* : *rx* : *tx* power ratios are 0.07:1:1.33:2.22 (Cisco 2004). These values must be compared with the ratios computed with the measured power consumption, which are 0.04:1:1.20:1.73 (see (Shih et al. 2002)). Measurements performed on other models of IEEE 802.11 wireless cards can be found in (Ebert et al. 2002; Feeney and Nilsson 2001). All the measurements reported in the literature have outlined an important point, that is, that any radio state transition comes at a significant energy cost (and time latency). This is especially true when the radio transits from the sleep (power down) to the idle (power up) state.

In this book, we model node energy consumption by using the *sleep* : *idle* : *rx* : *tx* power ratios. In other words, we are not interested in the absolute values of the power consumption, but on the relative values. In our simplified model, we assume that the radio is consuming conventional power 1 when the radio is idle, power 1.x when the radio is receiving a

Table 2.2 Nominal power consumption and transmit range of the CISCO IEEE 802.11 a/b/g wireless card. Power consumption is measured by the drain current, expressed in mA. In the table, the minimum value of the nominal range refers to the maximum data rate (54 Mbps), and the maximum value to the 6 Mbps data rate

	Power Idle (mA)	Power Tx (mA)	Power Rx (mA)
802.11 a	203	554	318
802.11 b	203	539	327
802.11 g	203	530	282

	Tx Range Indoor (m)	Tx Range Outdoor (m)
802.11 a	13–50	30–300
802.11 b/g	27–91	76–396

message, power $1.y$ when the radio is transmitting a message at full power, and power $0.z$ when the radio is in sleep mode (the actual values of x, y, and z depending on the specific card).

Before ending this section, we want to remark that the ratio $1.y$ used in our model refers to the relative power consumption of the radio when it transmits at maximum power. On the other hand, we shall see that topology control protocols are based on the ability of the wireless node to dynamically adjust its transmitting range. This feature is actually available on some commercial IEEE 802.11 cards, such as those produced by CISCO. For instance, the CISCO Aironet IEEE 802.11 a/b/g card can use transmit powers ranging from 1 mW to 100 mW. However, this value refers to the power consumption of the RF amplifier, which is only a part of the total power consumed by the wireless card. In fact, the card consumes significant energy also to power up the other analog and digital circuitry.

How to model power consumption when the radio is not transmitting at maximum power is not clear. Most of the approaches presented in the literature are concerned with the transmit power only, which is typically modeled using one of the formulas reported in Section 2.1. Unless otherwise specified, this book will also follow this simplistic energy model. In particular, we will use the following definition of energy cost:

Definition 2.3.1 (Energy cost) *Given a range assignment RA for a certain network $M_d = (N, L)$, the energy cost of RA is defined as*

$$c(RA) = \sum_{u \in N} RA(u)^\alpha,$$

where α is the distance-power gradient.

Note that the above definition of energy cost is coherent with our working assumption that the radio signal propagates according to the log-distance path model.

2.3.2 Sensor networks

In case of sensor networks, the task of providing a simple yet realistic energy model is relatively simpler, as compared to the case of ad hoc networks. In fact, sensor networks are

Table 2.3   Measured power consumption of a Rockwell's WINS sensor node

MCU Mode	Sensor Mode	Radio Mode	Total Power (mW)
On	On	Tx (power 36.3 mW)	1080.5
On	On	Tx (power 0.12 mW)	771.1
On	On	Rx	751.6
On	On	Idle	727.5
On	On	Sleep	416.3
On	On	Removed	383.3
Sleep	On	Removed	64.0

typically composed of homogeneous devices, which are usually very simple. Furthermore, since many sensor nodes have been designed in the research community, their features are very well known. As a result, several sets of energy consumption measurements of wireless sensor nodes have been reported in the literature (Raghunathan et al. 2002).

Table 2.3 reports the power dissipation of a Rockwell's WINS sensor node (Rock-wellScienceCenter 2004). The node is composed of three main components: the microcontroller unit (MCU), the sensing apparatus (sensor), and the wireless radio. If we consider the power consumption of the wireless radio only, we have the following *sleep : idle : rx : tx* ratios: 0.09:1:1.07:2.02. Note that these ratios are quite similar to the case of 802.11 wireless cards, except for a somewhat higher power consumption when the radio is transmitting at maximum power. When the transmit power is minimum (0.12 mW), the *idle : tx* ratio in the WINS sensor is 1.12. So, there is an almost twofold power consumption increase when varying the transmit power from the minimum to the maximum value. This means that varying the transmit power level has a considerable effect on the node's energy consumption.

2.4   Mobility Models

Node mobility is a prominent feature of ad hoc networks and, in some cases, also of WSNs. As a consequence, studying the performance of ad hoc/sensor networking protocols in the presence of mobility is a fundamental stage of the design process. Since real implementations of ad hoc/sensor networks are scarce, real-life movement patterns are very difficult to obtain, and the common approach is to use synthetic mobility models and simulation.

Mobility models for ad hoc/sensor networks should

- *resemble real-life movements*: Given the wide range of ad hoc and sensor network applications, the movement patterns to consider are numerous: they range from campuswide movement of students to vehicular motion in highways, from movement of groups of tourists in a urban scenario to rescue squads motion in disaster areas, and from sensors carried around by ocean flows to animal movement in animal tracking WSN applications. Providing a unique mobility model that resembles all these types of mobility is virtually impossible. However, a mobility model should be representative of at least one application scenario.

- *be simple enough for simulation/analysis*: Since mobility models are used in the simulation of ad hoc networks, the model should be simple enough to be integrated in the

simulator and to keep the simulation running time reasonable. Furthermore, using relatively simple mobility models eases the task of deriving meaningful analytical results
concerning fundamental network parameters in presence of mobility. In turn, these
results can be used to optimize the performance of ad hoc/sensor networking protocols.

Clearly, the two goals above are conflicting: the more realistic the model is, the more
the details that must be included in it, and the model complexity increases. Thus, a synthetic
mobility model should be a good compromise between representativeness and simplicity, that
is, it should consider the salient features of a certain movement pattern, while disregarding
secondary details.

In this section, we briefly describe the most important mobility models used in the
simulation of ad hoc/sensor networks. For a more detailed description of mobility models
for ad hoc networks, see (Bettstetter 2001a; Camp et al. 2002).

Random waypoint model.   This is by far the most commonly used mobility model for
ad hoc networks. One of the reasons for its popularity is the fact that it is implemented in
network simulations tools such as Ns2 (Ns2 2002) and GloMoSim (Zeng et al. 1998). The
Random Waypoint (RWP) model has been introduced in (Johnson and Maltz 1996) to study
the performance of the DSR routing protocol. In this model, each node chooses uniformly
at random a destination point (the 'waypoint') within the deployment region R, and moves
toward it along a straight line. Node velocity is chosen uniformly at random in the interval
$[v_{min}, v_{max}]$, where v_{min} and v_{max} are the minimum and maximum node velocities. When
the node arrives at destination, it remains stationary for a predefined pause time, and then
starts moving again according to the same pattern.

The RWP model is representative of an individual movement, obstacle-free scenario:
each node moves independent of each other (individual movement), and it can potentially
move in any subregion of R (obstacle-free). For instance, a similar type of mobility could
arise when users move in a large room, or in a open air, flat environment.

Given its popularity, RWP mobility has been deeply studied in the literature. In particular,
it has been recently discovered that the long-term node spatial distribution of RWP mobile
networks is concentrated in the center of the deployment region (border effect) (Bettstetter
and Krause 2001; Bettstetter et al. 2003; Blough et al. 2004), and that the average nodal
speed, defined as the average of the node velocities at a given instant of time, decreases over
time (Yoon et al. 2003). These observations have brought to the attention of the community
the fact that RWP mobile networks must be carefully simulated. In particular, network
performance should be evaluated only after a certain 'warm-up' period, which must be
long enough for the network to reach the node spatial and average velocity 'steady-state'
distribution.

The RWP model has also been generalized to slightly more realistic, though still simple,
models. For instance, in (Bettstetter et al. 2003) the RWP model is extended by allowing nodes to choose pause times from an arbitrary probability distribution. Furthermore, a
random fraction of the network nodes remains stationary for the entire simulation time.

Random direction model.   Similar to the RWP model, the random direction model resembles individual, obstacle-free movement. This model was created to maintain a uniform node
spatial distribution during the simulation time, thus avoiding the border effect typical of RWP
mobility.

In this model (Royer et al. 2001), any node chooses a direction uniformly at random in the interval $[0, 2\pi]$, and a random velocity in the interval $[v_{min}, v_{max}]$. Then, it starts moving in the selected direction with the chosen velocity. When the node reaches the boundary of R, it chooses a new direction and velocity, and so on.

Variants of this model have also been presented. In a first variant (Haas and Pearlman 1998; Pearlman et al. 2000b), a node is 'bounced back' when it reaches the boundary of the deployment region. In another (Bettstetter 2001b), the node moves for a random (exponentially distributed) time, and then it changes direction and velocity of movement.

Brownian-like motion. Contrary to the case of RWP and random direction mobility, which resemble intentional movement, the class of Brownian-like mobility models resembles nonintentional motion. For this reason, these models are sometimes called drunkardlike models.

In Brownian-like motion, the position of a node at a given time step depends (in a certain, probabilistic, way) on the node position at the previous step. In particular, no explicit modeling of movement direction and velocity is used in this model.

An example of Brownian-like motion is the model used in (Santi and Blough 2003). Mobility is modeled using three parameters: p_{stat}, p_{move} and m. The first parameter represents the probability that a node remains stationary for the entire simulation time. Parameter p_{move} is the probability that a node moves at a given time step. Parameter m models, to some extent, velocity: if a node is moving at step i, its position at step $i + 1$ is chosen uniformly at random in the square or side $2m$ centered at the current node position.

Map-based mobility. In all the models introduced so far, nodes are free to move within any subregion of the deployment region R. However, in many realistic scenarios, nodes are constrained to move within specified paths. This is the case, for instance, of cars moving on a freeway, or people moving on sidewalks, and so on. Map-based models have been used to model these situations.

The first step in the definition of map-based models is the map setup, that is, the definition of the paths within which nodes are allowed to move. Then, a certain number of nodes are randomly located on the paths, and they start moving according to scenario-specific rules.

An instance of map-based mobility is the Freeway mobility model (Bai et al. 2003), used to mimic the movement of cars on freeways. In this model, several freeways are located in the deployment region. Each freeway is composed of a varying number of lanes in both directions. Nodes are randomly located on a freeway, and they move with a random velocity, which is temporally dependent on its previous velocity. If two nodes on the same lane are within a certain minimum distance (safety distance), the velocity of the following node cannot exceed the velocity of the preceding one.

Another instance of map-based mobility is the Manhattan mobility model (Bai et al. 2003), which is used to emulate urban movement scenarios. First, a Manhattan-like map, composed of horizontal and vertical streets, is generated. Nodes can move along the streets in both directions. When a node arrives at an intersection, it randomly chooses whether to continue moving along the same direction, or to take a left or a right turn. Similar to the Freeway model, the velocity of a node at a certain instant of time depends on the node velocity at the previous time step.

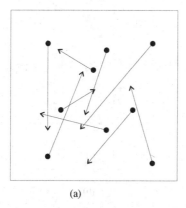
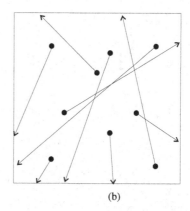

(a)   (b)

Figure 2.4 Examples of RWP (a) and random direction mobility (b). In case of RWP mobility, nodes tend to cross the center of the deployment region (border effect).

A third instance of map-based mobility is the Obstacle mobility model introduced in (Jardosh et al. 2003). In this model, a map is first generated by adding obstacles (buildings) to the environment. The obstacle-generation phase can be either random or based on real maps. Once the buildings are deployed, a number of pathways connecting different buildings are generated, and nodes are assumed to move along these pathways. An interesting feature of this model is that obstacles are accounted for also when simulating the radio signal propagation in the environment: in other words, it is assumed that the wireless signal is obstructed by the obstacles.

Group-based mobility. All the models described so far resemble individual mobility. However, in many situations, nodes are expected to move in groups (for instance, groups of tourists moving in a city). Group-based mobility has been introduced to model these situations.

In group-based models, a small subset of the network nodes is defined as the set of *group leaders*. The remaining nodes are randomly assigned to one of the leaders, thus forming groups. Initially, the leaders are randomly distributed in the deployment region R, and the members of each group are randomly located in the neighborhood of the leader. Then, the group leader moves according to one of the previous mobility models, such as RWP or random direction. The other group members 'follow' the leader, having a speed and direction that are a random perturbation of those of the leader. When two groups cross, any group member can leave its group and join the other with a certain probability. Group-based mobility models have been used in (Hong et al. 1999; Wang and Li 2002).

Examples of RWP and random direction mobility are shown in Figure 2.4, while Figure 2.5 reports examples of map-based mobility.

2.5 Asymptotic Notation

Before concluding this chapter, we recall the standard notation regarding the asymptotic behavior of functions.

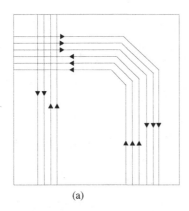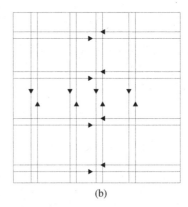

(a)   (b)

Figure 2.5 Examples of map-based mobility: the freeway model (a) and the Manhattan mobility model (b).

Let f and g be functions of a certain parameter x. We are interested in characterizing the asymptotic behavior of f and g as $x \to \infty$.

Definition 2.5.1 *We say that $f(x)$ has order at most $g(x)$, denoted as $f(x) \in O(g(x))$, if there exist constants c and x_0 such that, for any $x \geq x_0$, $f(x) \leq c \cdot g(x)$. We say that $f(x)$ has order at least $g(x)$, denoted as $f(x) \in \Omega(g(x))$, if $g(x) \in O(f(x))$. We say that $f(x)$ and $g(x)$ have the same order, denoted as $f(x) \in \Theta(g(x))$, if $f(x) \in O(g(x))$ and $f(x) \in \Omega(g(x))$. We will sometimes use the notation $f(x) \approx g(x)$ also to indicate that $f(x)$ and $g(x)$ have the same order.*

Definition 2.5.2 *We say that $f(x)$ is asymptotically smaller that $g(x)$, denoted as $f(x) \ll g(x)$, if $\lim_{x \to \infty} \frac{f(x)}{g(x)} = 0$. We say that $f(x)$ is asymptotically larger than $g(x)$, denoted as $f(x) \gg g(x)$, if $g(x) \ll f(x)$.*

3

Topology Control

3.1 Motivations for Topology Control

In Chapter 1, we have briefly described the many challenges that the ad hoc and sensor network designer must face. In this chapter, we start focusing on two of these challenges, which have motivated researchers to study the realm of topology control techniques.

3.1.1 Topology control and energy conservation

As outlined in Chapter 1, the efficient use of the scarce energy resources available to ad hoc and sensor network nodes is one of the fundamental tasks of the network designer. Since nodes consume a considerable amount of energy to transmit/receive messages (this is especially true in case of sensor networks), reducing the energy consumed for radio communications is an important issue.

Suppose node u must send a packet to node v, which is at distance d (see Figure 3.1). Node v is within u's transmitting range at maximum power, so direct communication between u and v is possible. However, there exists also a node w in the region C circumscribed by the circle of diameter d that intersects both u and v (see Figure 3.1). Since $\delta(u, w) = d_1 < d$ and $\delta(v, w) = d_2 < d$, sending the packet using w as a relay is also possible. Which of the two alternatives is more convenient from the energy-consumption point of view?

To answer this question, we must refer to specific wireless channel and energy-consumption models. For simplicity, assume the radio signal propagates according to the free space model and that we are interested in minimizing the transmit power only. With these assumptions, the power needed to send the message directly from u to v is proportional to d^2; in case the packet is relayed by node w, the total power consumption is proportional to $d_1^2 + d_2^2$.

Consider the triangle $\widehat{uwv}$, and let γ be the angle opposite to side uv. By elementary geometry, we have

$$d^2 = d_1^2 + d_2^2 - 2d_1d_2 \cos \gamma.$$

Topology Control in Wireless Ad Hoc and Sensor Networks   P. Santi
© 2005 John Wiley & Sons, Ltd

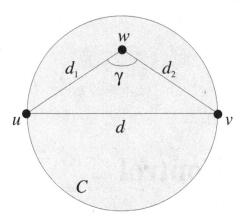

Figure 3.1 The case for multihop communication: node u must send a packet to v, which is at distance d; using the intermediate node w to relay u's packet is preferable from the energy consumption's point of view.

Since $w \in C$ implies that $\cos \gamma \leq 0$, we have that $d^2 \geq d_1^2 + d_2^2$. It follows that, *from the energy-consumption point of view, it is better to communicate using short, multihop paths between the sender and the receiver.*

The observation above gives the first argument in favor of topology control: instead of using a long, energy-inefficient edge, communication can take place along a multihop path composed of short edges that connects the two endpoints of the long edge. The goal of topology control is to identify and 'remove' these energy-inefficient edges from the communication graph.

3.1.2 Topology control and network capacity

Contrary to the case of wired point-to-point channels, wireless communications use a shared medium, the radio channel. The use of a shared communication medium implies that particular care must be paid to avoid that concurrent wireless transmissions corrupt each other.

A typical conflicting scenario is depicted in Figure 3.2: node u is transmitting a packet to node v using a certain transmit power P; at the same time, node w is sending a packet to node z using the same power P. Since $\delta(v, w) = d_2 < \delta(v, u) = d_1$, the power of the interfering signal received by v is higher than that of the intended transmission from u,[1] and the reception of the packet sent by u is corrupted.

Note that the amount of interference between concurrent transmissions is strictly related to the power used to transmit the messages. We clarify this important point with an example. Assume that node u must send a message to node v, which is experiencing a certain interference level I from other concurrent radio communications. For simplicity, we treat I as a received power level, and we assume that a packet sent to v can be correctly received only if the intensity of the received signal is at least $(1 + \eta)I$, for some positive η. If the current transmit power P used by u is such that the received power at v is below $(1 + \eta)I$,

---

[1]This is true independently of the deterministic path loss model considered. In case of probabilistic path loss models, this statement holds on the average.

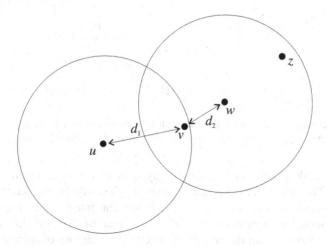

Figure 3.2 Conflicting wireless transmissions. The circles represent the radio coverage area with transmit power P.

we can ensure correct message reception by increasing the transmit power to a certain value $P' > P$ such that the received power at v is above $(1 + \eta)I$. This seems to indicate that increasing transmit power is a good choice to avoid packet drops due to interference. On the other hand, increasing the transmit power at u increases the level of interference experienced by the other nodes in u's surrounding. So, there is a trade-off between the 'local view' (u sending a packet to v) and the 'network view' (reduce the interference level in the whole network): in the former case, a high transmit power is desirable, while in the latter case, the transmit power should be as low as possible. The following question then arises: how should the transmit power be set, if the designer's goal is to maximize the network traffic carrying capacity?

In order to answer this question, we need an appropriate interference model. Maybe the simplest such model is the Protocol Model used in (Gupta and Kumar 2000) to derive upper and lower bounds on the capacity of ad hoc networks. In this model, the packet transmitted by a certain node u to node v is correctly received if

$$\delta(v, w) \geq (1 + \eta)\delta(u, v)$$

for any other node w that is transmitting simultaneously, where $\eta > 0$ is a constant that depends on the features of the wireless transceiver. Thus, when a certain node is receiving a packet, all the nodes in its *interference region* must remain silent in order for the packet to be correctly received. The interference region is a circle of radius $(1 + \eta)\delta(u, v)$ (the *interference range*) centered at the receiver. In a sense, the area of the interference region measures the amount of wireless medium consumed by a certain communication; since concurrent nonconflicting communications occur only outside each other interference region, this is also a measure of the overall network capacity.

Suppose node u must transmit a packet to node v, which is at distance d. Furthermore, assume there are intermediate nodes $w_1, \ldots, w_k$ between u and v and that $\delta(u, w_1) = \delta(w_1, w_2) = \cdots = \delta(w_k, v) = \frac{d}{k+1}$ (see Figure 3.3). From the network capacity point of

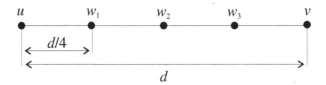

Figure 3.3 The case for multihop communication: node u must send a packet to v; using intermediate nodes $w_1, \ldots, w_3 = w_k$ is preferable from the network capacity point of view.

view, is it preferable to send the packet directly from u to v or to use the multihop path $w_1, w_2, \ldots, v$? This question can be easily answered by considering the interference range(s) in the two scenarios. In case of direct transmission, the interference range of node v is $(1 + \eta)d$, corresponding to an interference region of area $\pi d^2 (1 + \eta)^2$. In case of multihop transmission, we have to sum the area of the interference regions of each short, single-hop transmission. The interference region for any such transmission is $\pi \left(\frac{d}{k+1}\right)^2 (1 + \eta)^2$, and there are $k + 1$ regions to consider overall. Since, by Holder's inequality, we have

$$\sum_{i=1}^{k+1} \left(\frac{d}{k+1}\right)^2 = (k+1) \left(\frac{d}{k+1}\right)^2 < \left(\sum_{i=1}^{k+1} \frac{d}{k+1}\right)^2 = d^2,$$

we can conclude that, *from the network capacity point of view, it is better to communicate using short, multihop paths between the sender and the destination.*

The observation above is the other motivating reason for a careful design of the network topology: instead of using long edges in the communication graph, we can use a multihop path composed of shorter edges that connects the endpoints of the long edge. Thus, the maxpower communication graph, that is, the graph obtained when the nodes transmit at maximum power, can be properly pruned in order to maintain only 'capacity-efficient' edges. The goal of topology control techniques is to identify and prune such edges.

3.2 A Definition of Topology Control

In the previous section, we have presented at least two arguments in favor of a careful control of the network topology: reducing energy consumption and increasing network capacity. Although we have sometimes used the term 'topology control', a clear definition of it has not been introduced yet.

Quite informally, *topology control is the art of coordinating nodes' decisions regarding their transmitting ranges, in order to generate a network with the desired properties (e.g. connectivity) while reducing node energy consumption and/or increasing network capacity.*

While this definition is quite general, we believe that it captures the very distinguishing feature of topology control with respect to other techniques used to save energy and/or increase network capacity: *the networkwide perspective*. In other words, nodes make local choices (setting the transmit power level) with the goal of achieving a certain global, networkwide property. Thus, an energy-efficient design of the wireless transceiver cannot be classified as topology control because it has a nodewide perspective. The same applies to power-control techniques, whose goal is to optimize the choice of the transmit power level for a single wireless transmission, possibly along several hops; in this case, we have a channelwide perspective.

Note that our definition of topology control does not impose any constraint on the nature of the mechanism used to curb the network topology. So, both centralized and distributed techniques can be classified as topology control according to our definition.

Several authors consider as topology control techniques also mechanisms used to super-impose a network structure on an otherwise flat network organization. This is the case, for instance, of clustering algorithms, which organize the network into a set of clusters, which are used to ease the task of routing messages between nodes and/or to better balance the energy consumption in the network. Clustering techniques are more often used in the context of wireless sensor networks since these networks are composed of a very large number of nodes and a hierarchical organization of the network units might prove extremely useful.

In a typically clustering protocol, a distributed leader election algorithm is executed in each cluster, and cluster nodes elect one of them as the clusterhead. The election is based on criteria such as available energy, communication quality, and so on, or combination of them. Message routing is then performed on the basis of a two-level hierarchy: the message originating at a cluster node is destined to the clusterhead, which decides whether to forward the message to another clusterhead (intercluster communication) or to deliver the message directly to the destination (intracluster communication). The clusterhead might also perform other tasks such as coordinating sensor node sleeping times, aggregating the sensed data provided by the cluster nodes, and so on.

Although clustering protocols can be seen as a means of controlling the topology of the network by organizing its nodes into a multilevel hierarchy, a clustering algorithm does not fulfill our informal definition of topology control since typically the transmit power of the nodes is not modified. In other words, a clustering algorithm is concerned with hierarchically organizing the network units assuming the nodes' transmitting range is fixed, while a topology control protocol is concerned with how to modify the nodes' transmitting ranges in such a way that a communication graph with certain properties is generated.

3.3   A Taxonomy of Topology Control

As the informal definition of topology control introduced in the previous section outlines, many different techniques can be classified as topology control mechanisms. In this section, we try to organize these diverse approaches to the topology control problem in a coherent taxonomy. Our taxonomy of topology control techniques is depicted in Figure 3.4.

First, we distinguish between *homogeneous* CTR and *nonhomogeneous* topology control. In the former case, all the network nodes must use the same transmitting range r, and the topology control problem reduces to the simpler problem of determining the minimum value of r such that a certain networkwide property is satisfied. This value of r is known as the *critical transmitting range* (CTR), since using a range smaller than r would compromise the desired networkwide goal. In nonhomogeneous topology control, nodes are allowed to choose different transmitting ranges (subject to the condition that the chosen range does not exceed the maximum range).

The homogeneous case is by far the simplest formulation of the topology control prob-lem. Nevertheless, it has attracted the interest of many researchers in the field, probably

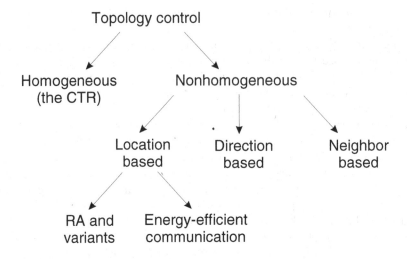

Figure 3.4 A taxonomy of topology control techniques.

because, owing to its simplicity, deriving clean theoretical results in this context is a challenging but feasible task. Chapters 4, 5, and 6 will be devoted to homogeneous topology control.

Nonhomogeneous topology control is classified into three categories, depending on the type of information that is used to compute the topology.

In location-based approaches, it is assumed that the most accurate information about node positions (the exact node location) is known. This information is either used by a centralized authority to compute a set of transmitting range assignments that optimizes a certain measure (this is the case of the Range Assignment problem and its variants), or it is exchanged between nodes and used to compute an 'almost optimal' topology in a fully distributed manner (this is the case of protocols for building energy-efficient topologies for unicast or broadcast communication). Typically, location-based approaches assume that network nodes, or at least a significant fraction of them, are equipped with GPS receivers. Location-based topology control techniques are described in Chapters 7 and 8 (centralized approach) and in Chapter 10 (distributed approach).

In direction-based approaches, it is assumed that nodes do not know their position but they can estimate the relative direction of their neighbors. This approach to topology control is discussed in Chapter 11.

In neighbor-based techniques, nodes are assumed to have access to a minimal amount of information regarding their neighbors, such as their ID, and to be able to order them according to some criterion (e.g., distance, or link quality). Neighbor-based techniques are probably the most suitable for application in mobile ad hoc networks, and are discussed in details in Chapter 12.

A final distinction is between *per-packet* and *periodical* topology control. In the former approach, every node maintains a list of efficient[2] neighbors and, for each such neighbor v, the transmit power to be used when sending packets to v. Thus, the choice of the transmit

---

[2]With efficient, we mean here either energy efficient, or capacity efficient, or both.

power to use is done on a per-packet basis: when the packet is destined to a certain neighbor v, the appropriate power $P(v)$ is set, and the packet is transmitted.

Per-packet topology control usually relies on quite accurate information on node locations, and it is typically applied in combination with location-based or direction-based topology control. A shortcoming of this technique is that it is rather demanding from a technological point of view, since it requires that the transmit power is changed very frequently (for an in-depth discussion of this issue, see Chapter 14). For this reason, simpler periodical techniques have been proposed. In this approach to topology control, every node maintains a list of efficient neighbors; however, differing from per-packet techniques, a node uses a single transmit power (the so-called *broadcast power*) to communicate with all the neighbors. This power can be intended as the higher of the transmit powers needed to reach the neighbors in the list. Periodically, the broadcast power level setting used by the node is updated, in response to node mobility and/or neighbor failures. As discussed in Chapter 13, periodical topology control is very suitable for application in mobile ad hoc networks.

3.4   Topology Control in the Protocol Stack

A final question is left: where should topology control mechanisms be placed in the ad hoc network protocol stack? Since there is no clear answer in the literature about this point, in what follows we describe our view, which is only one of the many possible solutions. In fact, the integration of topology control techniques in the protocol stack is one of the main open research areas in this field (see Chapter 15), and the best possible solution to this problem has not been identified yet.

In our view, topology control is an additional protocol layer positioned between the routing and MAC layer (see Figure 3.5).

3.4.1   Topology control and routing

The routing layer is responsible for finding and maintaining the routes between source/ destination pairs in the network: when node u has to send a message to node v, it invokes the routing protocol, which checks whether a (possibly multihop) route to v is known; if

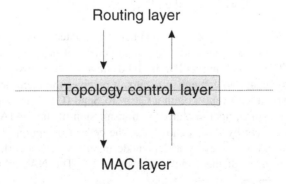

Figure 3.5   Topology control in the protocol stack.

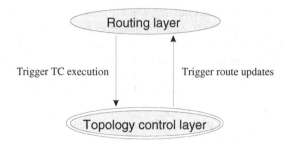

Figure 3.6 Interactions between topology control and routing.

not, it starts a route discovery phase, whose purpose is to identify a route to v; if no route to v is found, the communication is delayed or aborted.[3] The routing layer is also responsible for forwarding packets toward the destination at the intermediate nodes on the route.

The two-way interaction between the routing protocol and topology control is depicted in Figure 3.6. The topology control protocol, which creates and maintains the list of the immediate neighbors of a node, can trigger a route update phase in case it detects that the neighbor list is considerably changed. In fact, the many leave/join in the neighbor list are likely to indicate that many routes to faraway nodes are also changed. So, instead of passively waiting for the routing protocol to update each route separately, a route update phase can be triggered, leading to a faster response time to topology changes and to a reduced packet-loss rate. On the other hand, the routing layer can trigger the reexecution of the topology control protocol in case it detects many route breakages in the network, since this fact is probably indicative that the actual network topology has changed a lot since the last execution of topology control.

3.4.2 Topology control and MAC

The MAC (Medium Access Control) layer is responsible for regulating the access to the wireless, shared channel. Medium access control is of fundamental importance in ad hoc/sensor networks in order to reduce conflicts as much as possible, thus maintaining the network capacity to a reasonable level. To better describe the interaction between the MAC layer and topology control, we sketch the MAC protocol used in the IEEE 802.11 standard (IEEE 1999).

In 802.11, the access to the wireless channel is regulated through RTS/CTS message exchange. When node u wants to send a packet to node v, it first sends a Request To Send control message (RTS), containing its ID, the ID of node v, and the size of the data packet. If v is within u's range and no contention occurs, it receives the RTS message, and, in case communication is possible, it replies with a Clear To Send (CTS) message. Upon correctly receiving the CTS message, node u starts the transmission of the DATA packet, and waits for the ACK message sent by v to acknowledge the correct reception of the data.

In order to limit collisions, every 802.11 node maintains a Network Allocation Vector (NAV), which keeps trace of the ongoing transmissions. The NAV is updated each time

---

[3]We are considering here a reactive routing protocol, since there is wide agreement in the community that reactive routing performs better than proactive routing in ad hoc networks.

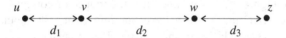

Figure 3.7 The importance of appropriately setting the transmit power levels.

a RTS, CTS, or ACK message is received by the node. Note that any node within u's and/or v's transmitting range overhears at least part of the RTS/CTS/DATA/ACK message exchange, thus obtaining at least partial information on the ongoing transmission.

As outlined, for instance, in (Jung and Vaidya 2002), using different transmit power levels can introduce additional opportunities for interference between nodes. On the other hand, using reduced transmit powers can also avoid interference. To clarify this point, consider the situation depicted in Figure 3.7. There are four nodes u, v, w, and z, with $\delta(u, v) = d_1 < d_2 = \delta(v, w)$ and $\delta(w, z) = d_3 < d_2$. Node u wants to send a packet to v, and node w wants to send a packet to z.

Assume all the nodes have the same transmit power, corresponding to transmitting range r, with $r > d_2 + \max\{d_1, d_3\}$. Then, the first between nodes v and z that sends the CTS message inhibits the other pair's transmission. In fact, nodes v and z are in each other's radio range, and overhearing a CTS from v (respectively, z) inhibits node z (respectively, v) from sending its own CTS. Thus, with this setting of the transmitting ranges, no collision occurs, but the two transmissions cannot be scheduled simultaneously.

Assume now that nodes u and v have radio range equal to r_1, with $r_1 = d_1 + \varepsilon < d_2$ and that nodes w and z have range r_2, with $r_2 > d_2$. In this situation, w and z cannot hear the RTS/CTS exchange between nodes u and v and they do not delay their data session. However, when node w transmits its packets, it causes interference at node v, which is within w's range. Thus, in this case, using different transmit powers *creates* an opportunity for interference.

Finally, assume nodes u and v have radio range r_1, and nodes w and z have range equal to r_3, with $r_3 = d_3 + \varepsilon < d_2$. With these settings of the radio ranges, the two transmissions can occur simultaneously, since node v is outside w's radio range and node z is outside u's radio range. Contrary to the example above, in this case, using different power levels *reduces* the opportunities for interference, leading to an increased network capacity.

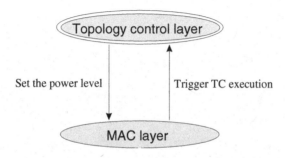

Figure 3.8 Interactions between topology control and MAC layer.

The example of Figure 3.7 has outlined the importance of correctly setting the transmit power levels at the MAC layer. We believe this important task should be performed by the topology control layer, which, having a networkwide perspective, can take the correct decisions about the node's transmitting range. On the other hand, the MAC layer can trigger reexecution of the topology control protocol in case it discovers new neighbor nodes. The MAC level can detect new neighbors by overhearing the network traffic and analyzing the message headers; this is by far the fastest way to discover new neighbors, and a proper interaction between MAC and topology control (which, we recall, is in charge of maintaining the list of efficient neighbors) ensures a quick response to changes in the network topology. The two-way interaction between topology control and the MAC layer is summarized in Figure 3.8.

Part II

The Critical Transmitting Range

4

The CTR for Connectivity: Stationary Networks

The simplest form of topology control considered in the literature is the characterization of the so-called *critical transmitting range* (CTR). In this version of topology control, all the network nodes are assumed to have the same transmitting range r, and the problem is to identify the minimum value of r (the critical range) such that certain networkwide properties are satisfied. The interest in finding the *minimum* value of r that guarantees certain properties is motivated by energy consumption and network capacity concerns (see Sections 3.1.1 and 3.1.2).

The most-studied version of the CTR problem in ad hoc and sensor networks is the characterization of the CTR for connectivity, that is, identifying the minimum value of r such that the resulting communication graph is connected.[1] The interest in characterizing the minimal conditions for connectivity lies in the fact that this is the most important network topological property. More formally, the problem can be stated as follows:

Definition 4.0.1 (CTR for connectivity) *Suppose n nodes are placed in a certain region $R = [0, l]^d$, with $d = 1, 2,$ or 3. Which is the minimum value of r such that the r-homogeneous range assignment is connecting?*

In the definition above, the deployment region is the d-dimensional cube with side l. This is only because most of the results presented in this and in the following chapters have been obtained for this shape of the deployment region. The definition of CTR for connectivity can be extended in a straightforward manner to deployment regions with arbitrary shape and size.

The assumption that all the nodes use the same transmitting range reflects all those situations in which transceivers use the same technology and no transmit power control. This is the case, for instance, for most of the 802.11 wireless cards currently on the market. In this scenario, using the same transmitting range for all the nodes is a reasonable choice,

---

[1] We recall that an undirected graph G is connected if and only if there exists at least one path connecting any two nodes in the graph.

and the only way to reduce energy consumption and increase capacity is to reduce r as much as possible (Narayanaswamy et al. 2002).

The following theorem shows that the CTR for connectivity equals the length of the longest edge of the Euclidean Minimum Spanning Tree (EMST) built on the network nodes (see Appendix A for the definition of EMST).

Theorem 4.0.2 *Let N be a set of n nodes placed in $R = [0, l]^d$, with $d = 1, 2,$ or 3. The CTR for connectivity r_C of the network composed of nodes in N equals the length of the longest edge of the EMST T built on the same set of nodes.*

Proof. Let e denote the longest edge in T, and let $l(e)$ denote its length. We first show that r_C cannot be larger than $l(e)$. This follows by observing that the $l(e)$-homogeneous range assignment produces a graph that contains T as a subgraph and that T is connected; by definition of CTR, we must have $r_C \leq l(e)$. Let us now prove that it cannot be that $r_C < l(e)$. Consider the sets of nodes corresponding to the two connected components T_1 and T_2 obtained from T by removing edge e (see Figure 4.1). By definition of EMST, edge e is the shortest edge connecting any pair (u, v) of nodes such that $u \in T_1$ and $v \in T_2$. Thus, any node in T_1 is at distance at least $l(e)$ from any node in T_2. This implies that setting the transmitting range to a value smaller than $l(e)$ would leave the communication graph disconnected, and the theorem is proved.

According to Theorem 4.0.2, computing the CTR[2] is equivalent to computing the EMST on the network nodes, and finding the longest edge in the EMST. Unfortunately, this way

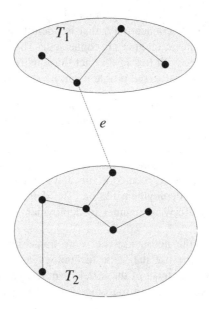

Figure 4.1   Connected components resulting from removing the longest edge e from the EMST.

---

[2]From now on, with CTR we mean CTR for connectivity (unless otherwise stated).

of calculating the CTR is not apt to distributed implementation, since building the EMST requires global knowledge (the exact positions of all the nodes in the network), which can be acquired in a distributed setting only by exchanging a considerable amount of messages. Furthermore, the requirement of knowing exact node positions is very strong: in fact, in many situations, node locations cannot be determined a priori (for instance, when sensors are dispersed on the field using a moving vehicle), and obtaining exact location information when nodes are already deployed is, in general, quite expensive (for instance, because many network nodes should be equipped with GPS receivers).

For the reasons described above, considerable attention has been devoted to characterizing the CTR in the presence of some form of uncertainty about node positions. If nodes' positions are not known, the minimum value of r ensuring connectivity in all possible cases is $r \approx l\sqrt{d}$, since nodes could be concentrated at the opposite corners of R. However, this scenario is overly pessimistic in many real-life situations. For this reason, a typical approach is to assume that nodes are distributed in R according to some probability density function $\mathcal{F}$, and to study the conditions for asymptotically almost sure connectivity.

Definition 4.0.3 (a.a.s. event) *Let E_k be a random event that depends on a certain parameter k. We say that E_k holds asymptotically almost surely (a.a.s.) or with high probability (w.h.p) if $\lim_{k \to \infty} P(E_k) = 1$.*

The probabilistic characterization of the CTR can be of great help in answering fundamental questions that arise at the network planning stage, such as: given a number n of nodes to be deployed in a certain region R, and given distribution $\mathcal{F}$, which resembles real-world node distribution, which is the minimum value $r_C(n, \mathcal{F})$ of the transmitting range that ensures connectivity with high probability? Conversely, given a transmitter technology (i.e. the value of r) and distribution $\mathcal{F}$, which is the minimal number $n_C(r, \mathcal{F})$ of nodes to be deployed in order to obtain a connected network with high probability?

The answer to the questions above depends on the shape of R and on the distribution $\mathcal{F}$ used to distribute nodes in R. In particular, we consider two probabilistic formulations of the CTR problem:

- *Fixed deployment region*: In this version of the problem, the side l of the deployment region R is fixed (e.g. R is the unit square), and the asymptotic value of the CTR as $n \to \infty$ is investigated. In principle, results obtained for this version of the problem can be applied only to dense networks. In fact, the value of the CTR is characterized as the node density $\frac{n}{l^d}$ grows to infinity, since l is an arbitrary constant.

- *Deployment region of increasing side*: In this version of the problem, the side l of the deployment region is a further model parameter, and the asymptotic value of the CTR as $l \to \infty$ is investigated. In this model, l can be seen as the independent variable, and both r and n are expressed as a function of l (and of the distribution $\mathcal{F}$). Since in this version of the problem the node density $\frac{n(l, \mathcal{F})}{l^d}$ can either converge to a constant $c \geq 0$ or diverge as $l \to \infty$, the theoretical results obtained using this model can be applied to networks with arbitrary density.

4.1   The CTR in Dense Networks

The CTR in dense networks can be characterized using results taken from a recent applied probability theory, the theory of *Geometric Random Graphs* (GRGs) (see Appendix B). Since the CTR equals the longest EMST edge, probabilistic solutions to the CTR problem in dense networks can be derived using results concerning the asymptotic distribution of the longest EMST edge.

The following theorem is proven in (Penrose 1997).

Theorem 4.1.1 (Penrose 1997) *Assume n points are distributed uniformly at random in the unit square* $[0, 1]^2$, *and let* M_n *be the random variable denoting the length of the longest MST edge built on the n nodes. Then,*

$$\lim_{n \to \infty} P[n\pi (M_n)^2 - \log n \leq \beta] = \frac{1}{\exp(e^{-\beta})},$$

for any $\beta \in \mathbf{R}$.

Corollary 4.1.2 *If R is the unit square and n nodes are distributed uniformly at random in R, then the CTR for connectivity is*

$$r_C = \sqrt{\frac{\log n + f(n)}{n\pi}},$$

where $f(n)$ *is an arbitrary function such that* $\lim_{n \to \infty} f(n) = +\infty$.

Proof. Let G_r denote the communication graph obtained when the transmitting range is set to r. Given the characterization of the CTR for connectivity of Theorem 4.0.2 and Theorem 4.1.1, G_r is a.a.s. connected if and only if

$$\lim_{n \to \infty} P\left[r \leq \sqrt{\frac{\log n + \beta}{n\pi}}\right] = 1. \tag{4.1}$$

It is immediate to see that Equality (4.1) is satisfied if and only if $\beta = f(n)$, for any function $f(n)$ such that $\lim_{n \to \infty} f(n) = +\infty$.

The CTR in case of three-dimensional networks can be derived by combining Theorem 1.4 of (Dette and Henze 1989) and Theorem 1.1 of (Penrose 1999a).

Theorem 4.1.3 *If R is the unit cube* $[0, 1]^3$ *and n nodes are distributed uniformly at random in R, then the CTR for connectivity is*

$$r_C = \sqrt[3]{\frac{\log n - \log \log n}{n\pi} + \frac{3}{2} \cdot \frac{1.41 + g(n)}{\pi n}},$$

where $g(n)$ *is an arbitrary function such that* $\lim_{n \to \infty} g(n) = +\infty$.

Note that, with respect to the case of two-dimensional networks and disregarding constants, the expression of the CTR in three-dimensional networks contains an additional· $\log \log n$ term. It is observed in (Dette and Henze 1989) that this term is due to the boundary effect (i.e. the presence of the border), which is asymptotically negligible in the two-dimensional case, while it is not negligible for three-dimensional networks.

In case of one-dimensional networks (nodes along a line), the CTR can be characterized by combining Theorem 1 of (Holst 1980), Theorem 2 of (Penrose 1997), and Theorem 2 of (Penrose 1999b).

Theorem 4.1.4 *If R is the segment of unit length [0, 1] and n nodes are distributed uniformly at random in R, then the CTR for connectivity is*

$$r_C = \frac{\log n}{n}.$$

We remark that the analysis of one-dimensional networks does have practical relevance, especially when modeling vehicular ad hoc networks (e.g. cars moving on a freeway).

Since the characterizations of the CTR stated in Corollary 4.1.2, Theorem 4.1.3, and Theorem 4.1.4 are asymptotic, an interesting question is: how fast is the convergence of the actual CTR to the asymptotic value? In other words, for which values of n are the values of the CTR predicted by our theorems accurate? This question can be answered by performing simulations and comparing the experimental results with those predicted by the asymptotic formulas.

Figure 4.2 depicts the rate of convergence of the actual CTR to the asymptotic value in case of two-dimensional networks, where the asymptotic value of the CTR is obtained by setting $f(n) = \log \log n$ in the formula given in Corollary 4.1.2 (this definition of $f(n)$ is sufficient in practice to achieve a value very close to 1 in the right-hand side of the formula of Theorem 4.1.1). The experimental value of the CTR is computed as follows: n nodes are distributed uniformly at random in $[0, 1]^2$, and the length of the longest EMST edge is recorded over a large set of experiments; the recorded values constitute the experimental

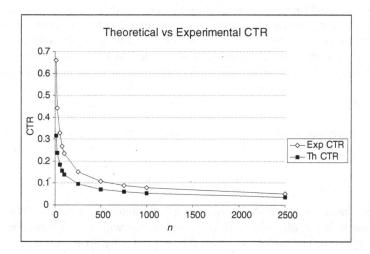

Figure 4.2   CTR for connectivity calculated according to Corollary 4.1.2 with $f(n) =$ $\log \log n$ (Th CTR), and experimental value of the CTR (Exp CTR) for different values of n.

Table 4.1   Values of the transmitting range yielding 99% of connected communication graphs for increasing network size. The table also reports the theoretical value of the CTR, calculated in accordance with Corollary 4.1.2

n	CTR–Th	CTR–99%	n	CTR–Th	CTR–99%
10	0.3160	0.6587	250	0.0959	0.1518
25	0.2364	0.4425	500	0.0716	0.1093
50	0.1833	0.3276	750	0.0601	0.0884
75	0.1566	0.2680	1000	0.0530	0.0773
100	0.1397	0.2349	2500	0.03547	0.0498

distribution of the longest EMST edge, which is used to derive the actual value of the CTR. The latter is defined as the .99 quantile of the experimental longest EMST edge distribution.[3] In other words, when the transmitting range is set to the critical value as defined above, the probability of generating a connected graph equals 0.99. The values of the CTR for connectivity obtained by simulation are reported in Table 4.1, which is based on the results presented in (Santi and Blough 2003).

As seen from Figure 4.2, the rate of convergence of the theoretical CTR value to the experimental one is low: with $n = 2500$, the relative difference between the theoretical and actual CTR is still in the order of 28%.

The giant component phenomenon.   An important phenomenon occurring in two- and three-dimensional GRGs is the so-called *giant component phenomenon*, which we now describe.

Consider a process in which all the network nodes have initially range $r = 0$, and then increase their transmitting range simultaneously. As the ranges are increased, new edges are added to the communication graph. We are interested in two particular instants of this process: the first instant at which the last isolated node disappears from the communication graph (i.e. the minimum node degree is at least one), and the first instant at which the communication graph becomes connected. Let us denote the ranges at these instants with r_I and r_C, respectively. It is clear that $r_I \leq r_C$. The following theorem, which is proven in (Penrose 1999a), states that for large values of n, $r_I = r_C$ a.a.s.

Theorem 4.1.5 (Penrose 1999a) *Assume n points are distributed in $R = [0, 1]^d$ according to the uniform distribution, with $d = 2, 3$. Let r_I and r_C be defined as above. Then,*

$$\lim_{n \to \infty} P[r_I = r_C] = 1.$$

An important consequence of Theorem 4.1.5 is the following: consider an instant of time corresponding to a range r_i that is close enough to r_C, with $r_i < r_C$; the communication graph at that time instant with high probability is formed by a large connected component (the *giant component*) plus few isolated nodes. Putting it another way, a relatively large connected component is formed quite soon in the increasing range process; as the range

---

[3] We recall that the q quantile of a series of data gives the point such that $100q$ percent of the data lie before.

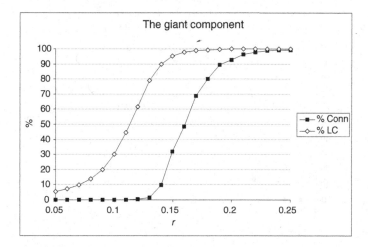

Figure 4.3 Percentage of nodes in the largest connected component (% LC) and percentage of fully connected graphs (% Conn) for different values of the transmitting range in two-dimensional networks with $n = 100$ network nodes.

increases further, additional nodes are added to this component, till the network becomes connected.

We remark that Theorem 4.1.5 has important practical implications. To better explain these implications, consider Figure 4.3, obtained through extensive simulation of two-dimensional networks with $n = 100$ nodes. The figure shows two curves: the higher curve refers to the average percentage of nodes belonging to the largest connected component of the communication graph; the lower curve refers to the percentage of fully connected communication graphs. Both plots are for increasing values of the transmitting range r.

The comparison of the two plots discloses the following important observation, which has its theoretical foundation in Theorem 4.1.5. If the network designer's goal is to produce a fully connected network with high probability, the transmitting range must be set to a relatively large value (approximately 0.23 when $n = 100$). However, if few isolated nodes can be tolerated, the required transmitting range can be considerably reduced: for instance, with $r = 0.14$, an average of 90% of the network nodes belong to the largest connected component, but the probability of generating a fully connected graph is only 0.1. This is because a giant component is formed quite soon in the increasing range process. Thus, tolerating few isolated nodes can have beneficial effect on energy consumption and network capacity.

In case of one-dimensional networks, the situation is quite different: as Figure 4.4 shows, the giant component phenomenon does not occur: the curves referring to the average largest connected component size and to the percentage of connected graphs are quite close to each other. This means that on the average, contrary to the case of two- and three-dimensional networks, connectivity occurs by joining several components of relatively large size. For instance, with $r = 0.07$ and $n = 100$, the percentage of connected networks is 94.7%, but the average size of the largest component in case of disconnected network is only $0.726n$.

The intuitive explanation of this different behavior, which is theoretically supported by the fact that Theorem 4.1.5 holds only for two- and three-dimensional networks, is the

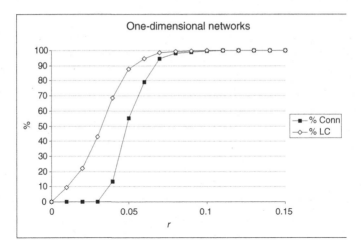

Figure 4.4 Percentage of nodes in the largest connected component (% LC) and percentage of fully connected graphs (% Conn) for different values of the transmitting range in one-dimensional networks with $n = 100$ network nodes.

following. In case of linear deployment region, an empty region of length at least r is sufficient to render the communication graph with range r disconnected (provided there is at least one node lying at both sides of the empty region). On the other hand, in case of two- and three-dimensional networks, a r-hole in one dimension is not sufficient to cause disconnection, because there could exist paths that 'go around the hole'. Then, network disconnection is caused by at least two- or three-dimensional holes, which occur with much smaller probability as compared to one-dimensional holes.

Before ending this section, we want to outline the similarities between the GRG model and the more traditional random graph model (Bollobás 1985), in which edges between arbitrary pair of nodes are randomly selected. In both the models, the graph with high probability becomes connected when the nodes have average degree in the order of $\log n$. Furthermore, the giant component phenomenon occurs in random graphs also, thus outlining another important similarity with two- and three-dimensional GRGs.

4.2   The CTR in Sparse Networks

A common assumption of the GRG model is that the node deployment region R is fixed (typically, it is a d-dimensional cube), and the asymptotic investigation is for increasing number of deployed nodes (i.e. for increasing density). Combining this observation with the fact that the rate of convergence of the actual CTR to the theoretical value of the CTR is quite low (see Figure 4.2), we can conclude that the results presented in the previous section in principle can be applied only to networks with very high node density. On the other hand, simple arguments based on interference considerations indicate that, in practice, node density cannot be too high.

To circumvent this problem, some authors suggested adding a further parameter to the model, the side l of the deployment region. In this model, l is the independent variable, and the asymptotic values of r and n (which can be seen as functions of l) yielding connectivity w.h.p. are investigated for $l \to \infty$. Differing from the GRG model, node density $\frac{n}{l^d}$ can either converge to 0, or to a constant $c > 0$, or diverge as $l \to \infty$, depending on the relative magnitude of n and l. Thus, theoretical results obtained in this model can be applied to both dense as well as sparse ad hoc networks.

Let us first consider one-dimensional networks. The following result, as well as the other results presented in this section, has been proven in (Santi and Blough 2003) by making use of the *occupancy theory* (see Appendix B), which is another applied probability theory used in the analysis of ad hoc network properties.

Theorem 4.2.1 (Santi and Blough 2003) *Assume n nodes, each with transmitting range r, are placed uniformly at random in $[0, l]$, and assume that $rn = kl \log l$, for some constant $k > 0$. Further, assume that $r = r(l) \ll l$ and $n = n(l) \gg 1$. If $k > 2$, or $k = 2$ and $r = r(l) \gg 1$, then the resulting communication graph is a.a.s. connected. If $k \leq (1 - \varepsilon)$ and $r = r(l) \in \Theta(l^\varepsilon)$ for some $0 < \varepsilon < 1$, then the communication graph is a.a.s. disconnected. If $r = r(l)$ is not of the form $\Theta(l^\varepsilon)$ but $rn \ll l \log l$, then the communication graph is a.a.s. disconnected.*

Corollary 4.2.2 *If $R = [0, l]$ and n nodes are distributed uniformly at random in R, the CTR for connectivity is*

$$r_C = k \frac{l \log l}{n},$$

where k is a constant with $1 \leq k \leq 2$.

As compared to Theorem 4.1.4, the statement of Theorem 4.2.1 is more involved, and contains several technical conditions. In particular, there are assumptions on the relative magnitudes of r and n when expressed as functions of the independent variable l, namely, $r = r(l) \ll l$ and $n = n(l) \gg 1$. Given the more general nature of this model as compared to the GRG model, these assumptions are necessary to investigate the asymptotic behavior of the CTR in a nontrivial setting. In fact, suppose $r \approx l$. In this case, each node has a direct connection to most of the other network nodes, and connectivity is ensured independent of n. On the other hand, if n would remain constant as l increases, the only way of obtaining a connected network would be to have $r \approx l$, which is also a trivial case.

It is interesting to compare Corollary 4.2.2 with the analogous theorem for dense networks. First of all, we observe that the characterization of the CTR in case of sparse networks is only partial since the exact value of the constant k is not known. By means of simulations, the authors of (Santi and Blough 2003) argue that k is probably 1, indicating a clear similarity with Theorem 4.1.4. Assuming $k = 1$, the only difference between the formulas presented in the two theorems is the 'geometric factor': while in case of fixed deployment region R the product $r_C n$ is proportional to $\log n$, in case of deployment region of side l, the product is proportional to $l \log l$. The l term can be interpreted as the scaling factor, while the $\log l$ term indicates the dependence of the CTR on a geometric parameter.

Figure 4.5 shows the rate of convergence of the actual CTR in one-dimensional networks to the asymptotic value as predicted by Corollary 4.2.2, where k is set to 1. As in the case of dense networks, the experimental value of the CTR is defined as the .99 quantile of the experimental longest MST edge distribution. In the experiments, the number n of nodes to distribute for a given value of l is set to $\lceil \sqrt{l} \rceil$. As seen from the figure, in this case,

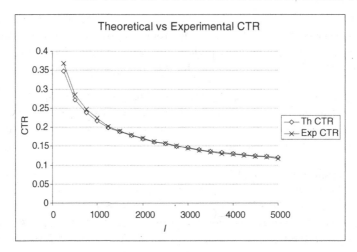

Figure 4.5 CTR for connectivity in one-dimensional networks calculated according to Corollary 4.2.2 with $k = 1$ (Th CTR), and experimental value of the CTR (Exp CTR) for increasing values of l. Parameter n is set to $\lceil \sqrt{l} \rceil$. The CTR reported on the y-axis is normalized with respect to l.

the asymptotic CTR formula of Corollary 4.2.2 is a very good approximation of the actual CTR for moderate to high values of l ($l = 1000$ and above). Note that these values of l correspond to values of n in the range 32–75. Thus, contrary to the case of dense networks, the formula of Corollary 4.2.2 is very accurate even for networks composed of few nodes.

In case of two- and three-dimensional networks, the characterization of the CTR proven in (Santi and Blough 2003) is weaker.

Theorem 4.2.3 (Santi and Blough 2003) *Assume n nodes, each with transmitting range r, are placed uniformly at random in $[0, l]^d$, with $d = 2, 3$ and assume that $r^d n = k l^d \log l$, for some constant $k > 0$. Further, assume that $r = r(l) \ll l$ and $n = n(l) \gg 1$. If $k > d k_d$, or $k = d k_d$ and $r = r(l) \gg 1$, then the resulting communication graph is a.a.s. connected, where $k_d = 2^d d^{d/2}$.*

Theorem 4.2.4 (Santi and Blough 2003) *Assume n nodes, each with transmitting range r, are placed uniformly at random in $[0, l]^d$, with $d = 2, 3$, and assume that $r = r(l) \ll l$ and $n = n(l) \gg 1$. If $r^d n \in O(l^d)$, then the resulting communication graph is not a.a.s. connected.*

Note the asymptotic gap between the necessary and sufficient condition for a.a.s. connectivity: it is known that $r^d n \in \Theta(l^d \log l)$ is sufficient for a.a.s. connectivity (Theorem 4.2.3) and that $r^d n \gg l^d$ is necessary for a.a.s. connectivity (Theorem 4.2.4). Thus, the CTR for connectivity r_C might be any function of the following type:

$$\frac{l^d f(l)}{n},$$

where $f(l)$ is a function such that $f(l) \in O(\log l)$ and $f(l) \gg 1$.

By means of extensive simulation, the authors of (Santi and Blough 2003) argue that $f(l) = \log l$ is also a necessary condition for a.a.s. connectivity. We then claim the following result, which is only partially proven.

Proposition 4.2.5 *If $R = [0, l]^d$, with $d = 2, 3$, and n nodes are distributed uniformly at random in R, the CTR for connectivity is*

$$r_C = k \frac{l^d \log l}{n},$$

where k is a constant with $0 \leq k \leq 2^d d^{d/2+1}$.

Let us finally comment about the giant component phenomenon in sparse ad hoc networks. Through simulations, it is observed in (Santi and Blough 2003) that the giant component phenomenon occurs in two- and three-dimensional networks, while it does not occur when nodes are located on a line. Although there is no formal proof of this fact, we can then conclude that sparse and dense ad hoc networks display the same behavior regarding the occurrence of the giant component.

4.3 The CTR with Different Deployment Region and Node Distribution

Characterizations of the CTR similar to those stated in the previous sections have been derived for different shapes of the deployment region R, and for different node distributions. In particular, Gupta and Kumar proved the same exact result as Corollary 4.1.2 when R is the disk of unit area (Gupta and Kumar 1998). The proof of Gupta and Kumar's result is based on the theory of *continuum percolation* (see Appendix B), which is another important applied probability theory used in the analysis of ad hoc network properties.

Other authors considered the case in which nodes are distributed according to a Poisson process of a given intensity λ. The CTR for Poisson distributed points on a line of length l is derived in (Piret 1991). A similar derivation of the CTR is obtained in (Dousse et al. 2002) when nodes are Poisson distributed on an unbounded one-dimensional region.

One observation regarding Poisson distribution is in order. With this type of distribution, the total number of deployed nodes is a random variable itself. In other words, with Poisson distribution, one is allowed to choose only the *expected* number of deployed nodes. For instance, if a Poisson process of intensity λ is used to distribute nodes on a line of length l, an average of $l\lambda$ nodes will be deployed. So, setting $\lambda = \frac{n}{l}$ generates a network with n nodes on the average. Given this observation, Poisson distribution is used whenever the exact number of network nodes is not known, but some information on the expected node density is available to the network designer.

Another distribution that has been considered in the literature is the Normal distribution. This distribution models those situations in which nodes are somewhat concentrated around a certain point. For instance, if an ad hoc network is used to provide wireless Internet access, it is reasonable to assume that nodes are concentrated around the access point. Another example in which assuming Normally distributed nodes is reasonable is when wireless sensors are deployed in groups using a vehicle (e.g. a helicopter): in this situation, node concentration around the release point is expected. The characterization of the CTR for connectivity in two- and three-dimensional networks with Normally distributed points is derived in (Penrose 1998).

Finally, we want to mention a recent result due to Penrose (Penrose 1999c), which characterizes the CTR for connectivity in case of *arbitrary* node distribution (provided certain technical conditions are satisfied). This important result, which is used in Chapter 5 to study the CTR in mobile ad hoc networks, essentially states that what determines the asymptotic behavior of the CTR is the minimal value of the probability density function $\mathcal{F}$ used to distribute nodes in the deployment region R.

4.4   Irregular Radio Coverage Area

As discussed in Section 2.2, the main limitation of the point graph model used to derive the results presented in this chapter is the assumption of regular radio coverage: for instance, in case of two-dimensional networks, the radio coverage region is assumed to be a disk of a certain radius centered at the transmitter. Given this weakness in the model, one might argue that the characterizations of the CTR introduced in the literature have scarce practical relevance. For this reason, some authors have recently investigated the conditions for a.a.s. connectivity in the presence of irregular radio coverage area. In this section, we discuss some interesting results presented in (Booth et al. 2003) and (Bettstetter 2004), which refer to two-dimensional ad hoc networks.

Consider a set of nodes located in the plane, and assume that nodes u and v are directly connected with a certain probability $g(\delta(u, v))$, where $\delta(u, v)$ is the distance between the two nodes. Typically, g is a decreasing function of the distance. However, this is not imposed in the model, which allows g to be an arbitrary function of the distance.

Since the radio connectivity is defined in probabilistic terms, the model above allows irregular radio coverage area. For instance, there could exist nodes u, v, w such that $\delta(u, v) < \delta(u, w)$, but only link (u, w) exists in the communication graph (see Figure 4.6). However, since the probability of having a link depends only on the distance between the nodes, the model can only represent situations in which the radio coverage area is rotary symmetric. For this reason, we call this model the *rotary symmetric connection model*.

The traditional point graph model can be expressed in the rotary symmetric connection model by defining

$$g_r(x) = \begin{cases} 1 & \text{if } x \leq r \\ 0 & \text{if } x > r \end{cases},\tag{4.2}$$

where r is the nodes' transmitting range. In this case, the (deterministic) radio coverage area is given by πr^2. In case of probabilistic wireless connections between nodes, the radio coverage area must be expressed in probabilistic terms. The natural way of doing this is by integrating the connectivity function $g()$ on $\mathbf{R}$. Formally, the radio coverage area $A(g)$ of the connectivity function g is defined as

$$A(g) = \int_{x \in \mathbf{R}} g(x) \, dx.$$

Note that the radio coverage area determines the expected number of neighbors. For instance, assuming that n nodes are distributed uniformly at random in the unit square, the expected number of neighbors of a certain node is given by $(n - 1)A(g)$.

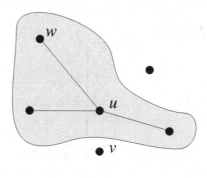

Figure 4.6 Example of radio coverage area (shaded region) in the rotary symmetric connection model. Node u is directly connected to node w, and it is not connected to the closer node v.

The authors of (Booth et al. 2003) investigate the connectivity properties of large ad hoc networks in the rotary symmetric connection model, considering different 'shapes' of the coverage region, with the constraint that the radio coverage area (and, hence, the expected number of neighbors) is the same. In particular, they consider Poisson distributed nodes, and characterize the minimal intensity of the Poisson process that enables the formation of an infinite connected component in the communication graph obtained by connecting nodes according to a certain function g, with $0 < A(g) < \infty$ (this condition is required to avoid trivial cases). Using the terminology of the continuum percolation theory, they analyze the *critical percolation density* λ_C, which is strictly related to the CTR for connectivity in the GRG model (see Appendix B).

Let g be any connectivity function such that $0 < A(g) < \infty$. Given parameter p with $0 < p < 1$, the *squashing transformation* g_p^{sq} of g is defined as follows:

$$g_p^{\mathrm{sq}}(x) = p \cdot g(\sqrt{p}x).$$

The connectivity function g_r defined in equation (4.2), and its squashing transformation $g_{r,1/2}$ of parameter $p = 1/2$ is reported in Figure 4.7. It is immediate to see that the squashing transformation preserves the radio coverage area, and, consequently, the expected number of neighbors of a node.

Intuitively, the squashing transformation of a certain connectivity function allows more faraway connections and reduces accordingly the probability of being connected to close nodes. In order words, direct connection to faraway nodes is traded off with reliable communication to close neighbors.

The following result, which was first stated in (Booth et al. 2003) and more formally proven in (Franceschetti et al. 2005), shows that long distance, unreliable links are at least as good as short distance, reliable links as far as network connectivity is concerned.

Theorem 4.4.1 *Let g be any connectivity function such that $0 < A(g) < \infty$ and let $\lambda_C(g)$ be the critical percolation density when nodes are connected according to g. For any $0 < p < 1$, we have*

$$\lambda_C(g) \geq \lambda_C(g_p^{\mathrm{sq}}).$$

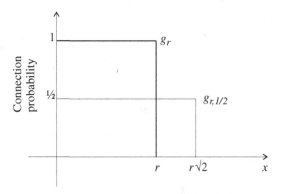

Figure 4.7 The connectivity function g_r, and its squashing transformation $g_{r,1/2}$ of parameter $p = 1/2$.

On the basis of theoretical argumentation and experimental results, the authors of (Booth et al. 2003) claim that a result similar to that of Theorem 4.4.1 holds also for a different type of transformation of g that preserves the radio coverage area: the shifting and squeezing transformation (see (Booth et al. 2003) and (Franceschetti et al. 2005) for details). They also claim that the circle of radius r is the shape that provides the highest critical percolation density, as compared to other shapes of the same area such as triangle, hexagon, and so on.

Summing up, we can conclude with the following fundamental statement:

Proposition 4.4.2 *Let $\lambda_C(r)$ denote the critical percolation density in the idealized point graph model, where the radio coverage area is a disk of radius r. Let $\lambda_c(g)$ denote the same density in the rotary symmetric connection model, where g is any connectivity function such that $A(g) = \pi r^2$. Then,*

$$\lambda_c(r) \geq \lambda_c(g).$$

We remark that the proposition above has not been formally proven yet, but it is supported by many theoretical and experimental evidences.

A similar conclusion to that stated in Proposition 4.4.2 is drawn in (Bettstetter 2004), in case the occurrence of wireless links obeys the log-normal shadowing model. We recall that in this model the path loss at distance d is modeled as a random variable with log-normal distribution centered around the mean value, which is derived using the classical log-distance path model (see Section 2.1). By using theoretical argumentation and extensive simulation, Bettstetter shows that the critical density for connectivity with deterministic radio coverage area (disk of a certain radius r) is at least as large as the same density considering shadowing effects.

The collection of the results presented in this section indicates that the characterization of the CTR based on the quite idealized point graph model can be considered as the worst-case scenario among all situations where the radio coverage area is the same. In other words, if the conditions for connectivity are met in the point graph model, then the same conditions are satisfied also in more realistic models that account for irregular coverage area, provided the wireless transmission footprint (and, hence, the expected number of neighbors) is the same. We can then conclude that the characterizations of the CTR for connectivity presented in Sections 4.1 and 4.2 do have practical significance.

5

The CTR for Connectivity: Mobile Networks

In the previous chapter, we have presented several characterizations of the CTR for connectivity in case of stationary networks. In this chapter, we analyze the effect of mobility on this important network parameter.

First, we have to agree on the definition of CTR in presence of mobility. Differing from the stationary case, the network topology changes with time because of node mobility. This implies that the CTR for connectivity also *changes with time*. Denoting with $t_1, t_2, \ldots, t_i, \ldots$ a sequence of time instants, we then have a sequence of values of the CTR for connectivity $r_1, r_2, \ldots, r_i, \ldots$, where r_i equals the length of the longest edge of the MST built on the n nodes at time t_i. Note that, in general, the values in the r_i sequence are neither increasing nor decreasing, that is, there could exist time instants i_1, i_2 with $i_1 < i_2$ such that $r_{i_1} < r_{i_2}$, and time instants i_3, i_4 with $i_3 < i_4$ such that $r_{i_3} > r_{i_4}$.

Several definitions of CTR for connectivity in mobile networks are possible. For instance, we could define the CTR as the maximum value of the r_is in the sequence of time instants corresponding to the network operational time. This is a very conservative definition of CTR, since it ensures that by setting the transmitting range to the critical value the resulting communication graph is connected during the entire network operational lifetime. However, this definition of mobile CTR in many situations might be too strong, because an occasional, extremely high value of one of the r_is would render the CTR very high as well. For this reason, in the literature, a definition of CTR for connectivity in presence of mobility that is based on a stochastic property of the mobile system has been introduced.

Definition 5.0.1 (Asymptotic node spatial distribution) *Assume n nodes are initially deployed in a certain region R according to some probability density function $\mathcal{F}$. After initial deployment, nodes start moving according to a certain mobility model $\mathcal{M}$. The asymptotic node spatial distribution generated by $\mathcal{M}$-like mobility with initial deployment $\mathcal{F}$ is the probability density function $\mathcal{F}_{\mathcal{M}}$ defined as*

$$\mathcal{F}_{\mathcal{M}} = \lim_{i \to \infty} \mathcal{F}_i, \qquad (5.1)$$

where $\mathcal{F}_i$ is the probability density function modeling node spatial distribution at time t_i. If the limit on the right-hand side of (5.1) does not exist, we say that mobility model $\mathcal{M}$ with initial deployment $\mathcal{F}$ does not stabilize.

Note the stochastic nature of this definition: it is assumed that initial node positions, as well as node positions at any time instant t_i, can be modeled by a certain probability density function that evolves with time. As discussed in Section 2.4, because of the lack of real movement patterns, a common approach in the evaluation of mobile ad hoc network properties is to use synthetic, stochasfic mobility models. Thus, our definition of asymptotic node spatial distribution is coherent with the stochastic nature of mobility models for ad hoc networks.

We are now ready to define the CTR in presence of mobility.

Definition 5.0.2 (Mobile CTR) *Assume n nodes are initially deployed in a certain region R according to some probability density function $\mathcal{F}$. After initial deployment, nodes start moving according to a certain mobility model $\mathcal{M}$. The CTR for connectivity in $\mathcal{M}$-mobile networks with initial deployment $\mathcal{F}$ is defined as the minimum value of the transmitting range that ensures a.a.s connectivity under the assumption that n nodes are distributed in R with density $\mathcal{F}_\mathcal{M}$, where $\mathcal{F}_\mathcal{M}$ is the asymptotic node spatial distribution generated by $\mathcal{M}$-like mobility with initial deployment $\mathcal{F}$.*

Implicit in the definition above is the fact that the mobility model stabilizes. This is actually the case of most of the models considered in the literature (for instance, all the models described in Section 2.4). In case of unstable mobility patterns, a different definition of CTR (such as the maximum of the r_is sequence) should be used.

By defining the CTR in presence of mobility as above, we can prove an ergodic property of certain mobile networks. We recall that a stochastic process composed of a sequence of random variables $r_1, r_2, \ldots, r_i, \ldots$ (in our case, the sequence of the longest MST edge lengths) is *ergodic* if sampling from the sequence of random variables is statistically equivalent to repeatedly sampling from a certain, fixed random variable (in our case, the length of the longest MST edge computed when nodes are distributed according to $\mathcal{F}_\mathcal{M}$).

Theorem 5.0.3 *Let $\mathcal{M}$ be a stable and c-independent mobility model, that is, a model such that node positions at time t_{i+c} are independent of node positions at time t_i, for some constant $c > 0$. Then, a network with $\mathcal{M}$-like mobility is ergodic with respect to the CTR for connectivity.*

Proof. Consider an $\mathcal{M}$-mobile network. Let $r_1, r_2, \ldots, r_i, \ldots$ denote the sequence of random variables corresponding to the critical range for connectivity computed at time $t_1, t_2, \ldots, t_i, \ldots$. By hypothesis, $\mathcal{M}$ is stable, that is, there exists a probability density function $\mathcal{F}_\mathcal{M}$ such that, for i sufficiently large, r_i has the same distribution as that of random variable $\bar{r}$, where $\bar{r}$ denotes the length of the longest MST edge when nodes are distributed according to $\mathcal{F}_\mathcal{M}$.

Let us consider two consecutive random variables r_i and r_{i+1} in the sequence. In general, r_{i+1} is not independent of r_i, since node positions at time t_{i+1} might depend on node positions at the previous step (for instance, because nodes are moving along a certain trajectory). However, by hypothesis, there exists a constant $c > 0$ such that, for any i sufficiently large, the node positions at time t_{i+c} are independent of node positions at time t_i. Thus, variables r_i and r_{i+c} are independent, and sampling from r_i and subsequently from r_{i+c} is statistically equivalent to sampling twice from $\bar{r}$ (this is true because also random variable r_{i+c} has the same distribution as $\bar{r}$). Given this observation, we can subdivide the original sequence of random variables $S = \{r_i, r_{i+1}, \ldots, r_{i+c}, r_{i+1+c}, \ldots\}$ into c subsequences $S_0 = \{r_i, r_{i+c}, r_{i+2c}, \ldots\}$, $S_1 = \{r_{i+1}, r_{i+1+c}, r_{i+1+2c}, \ldots\}$, $\ldots$. For any such subsequence S_j, successively sampling from S_j is statistically equivalent to repeatedly sampling from $\bar{r}$ (this is because the random variables in S_j are independent). Since any random variable in the original sequence S belongs to one and only one of the S_js, it follows that successively sampling from S is statistically equivalent to repeatedly sampling from $\bar{r}$, and the theorem is proven.

Intuitively, ergodicity adds a temporal dimension to our definition of CTR in presence of mobility. To better understand this point, assume that the transmitting range is set to a value r such that the probability of generating a connected graph when nodes are distributed according to $\mathcal{F}_M$ is 0.99, and assume $\mathcal{M}$ satisfies the hypotheses of Theorem 5.0.3. By ergodicity, we can state that, on the average, 99% of the values observed in the sequence S of the longest MST edge lengths is below r. This means that if we observe the mobile network for a sufficiently long period of time then the fraction of time in which the network is connected approaches 0.99. So, by observing the network behavior when nodes are distributed according to the asymptotic node spatial distribution generated by $\mathcal{M}$-like mobility, we can obtain information on the dynamic behavior of the network when nodes move.

The discussion above has outlined that, assuming $\mathcal{M}$ is a stable mobility model, the problem of characterizing the CTR in presence of $\mathcal{M}$-like mobility can be reduced to studying the CTR under the assumption that nodes are distributed according to a certain distribution $\mathcal{F}_M$.

The first observation is that if a certain mobility model $\mathcal{M}$ generates a uniform asymptotic node spatial distribution (i.e. $\mathcal{F}_M$ is the uniform distribution) then the results on the CTR presented in the previous chapter can be directly applied to $\mathcal{M}$-mobile networks. An example of such mobility model is Brownian-like mobility (see Section 2.4 for the definition of Brownian-like motion): in (Blough et al. 2003b), it is shown through simulation that this mobility model generates a uniform long-term node spatial distribution.

In the next section, we consider the case of RWP mobility, which is the only mobility model for which the asymptotic node spatial distribution has been derived.

5.1   The CTR in RWP Mobile Networks

In this section, we characterize the CTR for connectivity in case of RWP mobility, which is by far the most popular mobility model used in the simulation of ad hoc networks.

It is known that the asymptotic node spatial distribution generated by RWP mobility is not uniform but is somewhat concentrated in the center of the deployment region (Bettstetter and Krause 2001; Blough et al. 2004). This phenomenon, which is called the *border effect*, is due to the fact that the waypoints (i.e. the destinations of a movement) in the RWP model are selected uniformly at random in a bounded deployment region R. To better understand this point, consider a RWP mobile node u, and assume that node u is currently

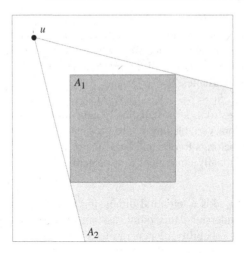

Figure 5.1 The border effect in RWP mobile networks: when a node is resting close to the border, it is likely that the trajectory to the next waypoint crosses the center of the deployment region (dark shaded area). In the figure, the probability that the trajectory of node u to the next waypoint intersects A_1 equals the sum of the areas of A_1 and A_2 (we are assuming $R = [0, 1]^2$).

resting at a waypoint that is close to the border of R (see Figure 5.1). Since the next waypoint is chosen uniformly at random in R, it is very likely that the trajectory connecting node u with its next waypoint will cross the center of R. So, the probability of finding a mobile node close to the center of R is higher than the probability of finding the node on the boundary. This means that mobile nodes contribute a nonuniform component to the asymptotic node spatial distribution generated by RWP mobility, which we denote by $\mathcal{F}_m$ (m stands for 'mobile'). On the other hand, a node resting at a waypoint contributes a uniform component $\mathcal{F}_u$ to the asymptotic RWP distribution, since the waypoints are chosen uniformly at random in R. Then, the asymptotic node spatial distribution generated by RWP mobility, denoted by $\mathcal{F}_{RWP}$, is given by $\mathcal{F}_{RWP} = \mathcal{F}_m + \mathcal{F}_u$, which is nonuniform. The amount of this nonuniformity (and, hence, the intensity of the border effect) depends on the relative strength of the two components of $\mathcal{F}_{RWP}$. It is easy to see that a longer pause time strengthens $\mathcal{F}_u$, since the nodes remain stationary for a longer time. Conversely, $\mathcal{F}_m$ is maximal when the pause time is 0 because, in this case, nodes are constantly moving.

The informal argument above is theoretically supported by the following theorem proven in (Bettstetter et al. 2003), which derives a very good approximation of $\mathcal{F}_{RWP}$ when nodes move in $R = [0, 1]^2$.

Theorem 5.1.1 (Bettstetter et al. 2003) *The asymptotic spatial density function of a node moving in $R = [0, 1]^2$ according to the RWP model with pause time t_p and velocity v is closely approximated by*

$$\mathcal{F}_{RWP}(x, y) = \begin{cases} P_{pause} + (1 - P_{pause})\mathcal{F}_m(x, y) & \text{if}(x, y) \in [0, 1]^2 \\ 0 & \text{otherwise} \end{cases},$$

where $P_{\text{pause}} = \dfrac{t_p}{t_p + \frac{0.521405}{v}}$ *and*

$$\mathcal{F}_m(x, y) = \begin{cases} 0 & \text{if}(x = 0) \text{ or } (y = 0) \\ \mathcal{F}_{\mathcal{R}}(x, y) & \text{otherwise} \end{cases}.$$

The expression of $\mathcal{F}_{\mathcal{R}}(x, y)$ *is the following:*

$$\mathcal{F}_{\mathcal{R}}(x, y) =$$

$$6y + \frac{3}{4}(1 - 2x + 2x^2)\left(\frac{y}{y-1} + \frac{y^2}{(x-1)x}\right) +$$

$$\frac{3}{2}\left((2x - 1)y(1 + y)\log\left(\frac{1-x}{x}\right) + y(1 - 2x + 2x^2 + y)\log\left(\frac{1-y}{y}\right)\right).$$

We remark that the expression of $\mathcal{F}_m(x, y)$ above is valid only for $(x, y) \in \mathcal{R} = \{(x, y) \in [0, 1]^2 \mid (x \geq y) \wedge (x \leq 1/2)\}$. The expression of $\mathcal{F}_m(x, y)$ on the remainder of $[0, 1]^2$ can be easily obtained observing that by symmetry we have $\mathcal{F}_m(x, y) = \mathcal{F}_m(y, x) = \mathcal{F}_m(1 - x, y) = \mathcal{F}_m(x, 1 - y)$.

The 3D plot of $\mathcal{F}_{\text{RWP}}$ for different values of the pause time is reported in Figure 5.2: as predicted by Theorem 5.1.1, longer pause times generate a flatter probability density function.

The CTR in presence of RWP mobility can be characterized by using the following result of the GRG theory, which is due to Penrose (Penrose 1999c).

Theorem 5.1.2 (Penrose 1999c) *Assume n nodes are distributed independently at random in* $\mathbf{R}^2$ *according to a common probability density function* $\mathcal{F}$, *having connected and compact support* Ω *with smooth boundary* $\partial\Omega$. *Further, assume that* $\mathcal{F}$ *is continuous on* $\partial\Omega$. *Let* M_n *denote the length of the longest MST edge built on the n points. Then,*

$$\lim_{n \to \infty} \frac{n\pi (M_n)^2}{\log n} = \frac{1}{\min_{\Omega} \mathcal{F}}, \tag{5.2}$$

almost surely.

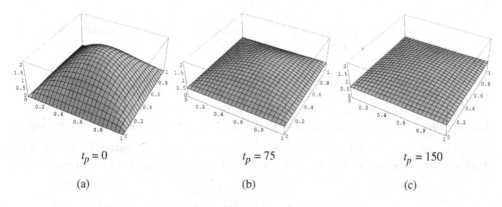

$t_p = 0$ $t_p = 75$ $t_p = 150$

(a)   (b)   (c)

Figure 5.2 3D plot of $\mathcal{F}_{\text{RWP}}$ for three different values of t_p: $t_p = 0$ (a), $t_p = 75$ time steps (b), and $t_p = 150$ time steps (c). Velocity v is set to 0.01 units per time step.

We recall that the support Ω of a probability density function is the set of points in which it has nonzero value, and that the boundary $\partial\Omega$ is smooth if and only if it is twice differentiable.

Informally speaking, Theorem 5.1.2 states that the asymptotic behavior of the CTR for connectivity with arbitrary density $\mathcal{F}$ depends only on the minimum value of $\mathcal{F}$ in its support. In case $\min_\Omega \mathcal{F} = 0$, the limit in equation (5.2) must be intended as $+\infty$.

In order to apply Theorem 5.1.2 to $\mathcal{F}_{\text{RWP}}$, we have to check that all the conditions of the theorem are satisfied. It is immediate to see that $R = [0, 1]^2$, the support of $\mathcal{F}_{\text{RWP}}$, is connected and compact. However, the boundary ∂R of R is not smooth because of the presence of the corners. This problem can be circumvented by using the 'corner-rounding' technique described in (Santi 2005). Thus, we are in the hypotheses of Theorem 5.1.2, and the only thing left to do to characterize the CTR is to determine the minimum value of $\mathcal{F}_{\text{RWP}}$ in R. This can be easily done, given the expression of $\mathcal{F}_{\text{RWP}}$ introduced in Theorem 5.1.1.

Corollary 5.1.3 *Let $\mathcal{F}_{RWP}^{t_p}$ denote the asymptotic node spatial density generated by RWP mobile networks with pause time t_p and velocity v. The minimum value of $\mathcal{F}_{RWP}^{t_p}$ is achieved on ∂R, and it equals $P_{pause} = \frac{t_p}{t_p + \frac{0.521405}{v}}$. When $t_p \to \infty$, $\mathcal{F}_{RWP}^{t_p}$ becomes the uniform distribution on $[0, 1]^2$, and $\min_R \mathcal{F}_{RWP}^{\infty} = 1$.*

We are now ready to characterize the CTR in presence of RWP mobility.

Theorem 5.1.4 (Santi (2005)) *If $R = [0, 1]^2$ and n nodes move in R according to the RWP mobility model with pause time t_p and velocity v, then the CTR for connectivity is*

$$r_{\text{RWP}}^{t_p} = \frac{1}{P_{\text{pause}}} \sqrt{\frac{\log n}{\pi n}} = \frac{t_p + \frac{0.521405}{v}}{t_p} \sqrt{\frac{\log n}{\pi n}}$$

if $t_p > 0$. When $t_p = 0$, we have

$$r_{\text{RWP}}^0 \gg \sqrt{\frac{\log n}{n}}$$

a.a.s.

Note that the CTR in presence of RWP mobility is always larger than the CTR in case of uniform node distribution since $1/P_{\text{pause}}$ is larger than 1 for any value of t_p. For instance, with $t_p = 75$ and $v = 0.01$, we have $1/P_{\text{pause}} = 1.69485$. Clearly, a longer pause time results in a more uniform node distribution and, consequently, in a smaller value of the CTR. For instance, with $t_p = 150$, we have $1/P_{\text{pause}} = 1.34743$.

Note also the asymptotic gap of the CTR in the most extreme case of RWP mobility, that is, when $t_p = 0$: in this case, for any constant $c > 0$, setting the transmitting range to $c\sqrt{\frac{\log n}{n}}$ is not sufficient for achieving a.a.s. connectivity. The exact value of the CTR with RWP mobility when $t_p = 0$ is not known to date. In (Santi 2005), it is conjectured that

$$r_{\text{RWP}}^0 \approx \frac{1}{4} \log n \sqrt{\frac{\log n}{\pi n}}.$$

This formula is supported by experimental evidence.

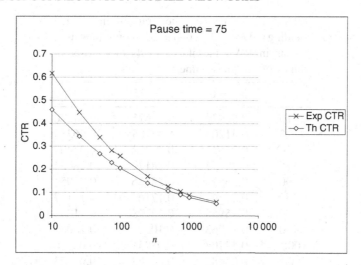

Figure 5.3 CTR for connectivity in case of RWP mobility with $t_p = 75$ and $v = 0.01$, for increasing values of n. The lower plot (ThCTR) refers to the asymptotic value, calculated in accordance with Theorem 5.1.4. The upper plot (ExpCTR) is obtained from the experimental CTR distribution generated by the simulations.

Figure 5.3 shows the rate of convergence of the actual CTR for connectivity to the asymptotic value stated in Theorem 5.1.4 in case of RWP mobility with $t_p = 75$. The actual CTR value is computed as follows. Initially, n nodes are distributed uniformly at random in $R = [0, 1]^2$. Then, they start moving according to the RWP mobility model. After a large number of mobility steps (1000 in our experiments), nodes' positions are recorded, and utilized to generate the experimental distribution of the longest MST edge length in case of mobility. As in the case of stationary networks, the experimental CTR value is defined as the 0.99 quantile of this distribution.

From the Figure, it is seen that the formula of Theorem 5.1.4 is quite accurate only for large values of n ($n = 1000$ and above). The experimental value of the CTR for RWP mobile networks with different values of the pause time is reported in Table 5.1.

Before concluding this section, we prove that the RWP mobility model satisfies the conditions for ergodicity.

Theorem 5.1.5 *A network with RWP mobility is ergodic with respect to the CTR for connectivity.*

Proof. In order to prove the theorem, we have to show that the RWP mobility model is stable and c-independent, for some constant $c > 0$. The first property is an immediate consequence of Theorem 5.1.1. As for the second, consider an arbitrary time instant i. We have to determine a certain value $c > 0$ such that the positions of all the nodes at time $i + c$ are independent of node positions at time i. Let us define a movement epoch as the time needed for a node just arrived at a waypoint to reach the next waypoint. In other words, a movement epoch is composed of the pause time plus the travel time between two consecutive waypoints. Since the length of the trajectory and node velocity are in general

Table 5.1 Values of the transmitting range
yielding 99% of connected communication
graphs in RWP mobile networks, for different
values of the pause time t_p

n	$t_p = 0$	$t_p = 75$	$t_p = 150$
10	0.56423	0.61625	0.64226
25	0.41203	0.44705	0.46285
50	0.33644	0.33892	0.34404
75	0.29454	0.28179	0.28054
100	0.26526	0.25736	0.2395
250	0.19761	0.17163	0.17117
500	0.15955	0.12728	0.1134
750	0.13963	0.10507	0.10086
1000	0.12708	0.08931	0.08416
2500	0.09482	0.05963	0.05473

random variables, the duration of a movement epoch is also a random variable. Indeed, we have a sequence of random variables representing the duration of the various epochs that constitute the movement trace of a node. We denote these variables with $E_{u,j}$, where u is the node to which the variable is referred and j denotes the jth epoch of node u. By definition of RWP mobility, node u's position at time $i + c$ is independent of its position at time i if and only if c is larger than $E_{u,j} + E_{u,j+1}$, where j is the index of the epoch occurring at time i. In words, the node must conclude the current and the next epoch before its position is independent of the position at time i. Note that it is not enough for the node to terminate the current epoch, since a node which is traveling at time i is on its trajectory to a certain waypoint $W_{u,j}$, which is also the starting point of the next trajectory. However, after the node has reached the next waypoint, the conditions for independence are satisfied. So, proving the theorem reduces to proving that there exists constant $c > 0$ such that $E_{u,j} + E_{u,j+1} \leq c$, for any $j \geq 0$ and for any node u. This is accomplished by setting $c = 2\frac{\sqrt{2}}{v_{min}}$. In fact, the maximum length of a linear trajectory in $R = [0, 1]^2$ is $\sqrt{2}$, and node velocity in the RWP model is at least $v_{min} > 0$. Note that, by setting $c = 2\frac{\sqrt{2}}{v_{min}}$, we ensure that the positions of all the nodes at time $i + c$ are independent of their positions at time i. This follows from the fact that inequality $E_{u,j} + E_{u,j+1} \leq c$ is satisfied for any epoch and for any node.

Given the ergodicity property of Theorem 5.1.5, the CTR values reported in Table 5.1 can be interpreted as the values of the transmitting range such that the RWP mobile network is connected for 99% of its operational time.

5.2 The CTR with Bounded, Obstacle-free Mobility

In this Section, we show that Penrose's characterization of the longest MST edge length with arbitrary node distribution (Theorem 5.1.2) can be used to partially characterize the

CTR of other types of mobile networks. In particular, we consider *bounded, obstacle-free* mobility models, which are defined as follows.

Definition 5.2.1 (Bounded, obstacle-free mobility) *Let $\mathcal{M}$ be an arbitrary mobility model and let $\mathcal{F}_\mathcal{M}$ be its asymptotic node spatial distribution (under the assumption that nodes are initially deployed according to a certain probability density function $\mathcal{F}$). $\mathcal{M}$ is bounded if and only if there exists a bounded region R such that the support of $\mathcal{F}_\mathcal{M}$ is contained in R. Furthermore, $\mathcal{M}$ is obstacle free if the support of $\mathcal{F}_\mathcal{M}$ contains $R - \partial R$.*

In words, a mobility model is bounded if there exists a bounded region R such that nodes are allowed to move only within R, while it is obstacle free if the probability of finding a mobile node in any subregion of R (excluding the border) is greater than 0.

Note that most of the mobility models used in the simulation of ad hoc and sensor networks are bounded and obstacle free; this is the case, for instance, of the random direction model, of Brownian-like mobility models, and of most group-based mobility models.

Theorem 5.2.2 (Santi 2005) *Let $\mathcal{M}$ be an arbitrary mobility model that is bounded within $R = [0, 1]^2$ and obstacle free. Furthermore, assume that $\mathcal{F}_\mathcal{M}$ is continuous on ∂R, and $\min_R \mathcal{F}_\mathcal{M} > 0$. The CTR for connectivity of an ad hoc network with $\mathcal{M}$-like mobility is*

$$r_\mathcal{M} = c\sqrt{\frac{\log n}{\pi n}},$$

for some constant $c \geq 1$.

Since in case of uniform node distribution the constant c in the expression of the CTR above equals 1, Theorem 5.2.2 can be interpreted as follows: *every bounded and obstacle-free type of node mobility is detrimental for network connectivity*, since the CTR for connectivity can only increase with respect to the case of uniformly distributed nodes. However, we remark that this result is asymptotic, that is, it holds for networks composed of a large number of nodes. If the network is composed of a relatively small number of nodes (say, in the order of 100) the situation might even be reversed (see (Santi 2005) for some simulation results that support this observation).

The final comment is regarding the occurrence of the giant component phenomenon in case of mobile networks. By combining Theorem 1.1 of (Penrose 1999b) and Theorem 1.1 of (Penrose 1999c), it can be formally proven that the giant component phenomenon occurs in any (two- or three-dimensional) bounded, obstacle-free mobile network. This fact is also supported by the simulation results presented in (Santi and Blough 2002), which refer to the case of RWP and Brownian-like mobile networks. Thus, connectivity can be traded off with energy saving and/or capacity increase also in presence of certain types of node mobility.

6

Other Characterizations of the CTR

In the previous chapter, we have presented several characterizations of the critical value of the transmitting range needed for guaranteeing the most important network property, that is, connectivity. In this chapter, we consider characterizations of the critical value of the range for other important network properties, such as k-connectivity, connectivity with Bernoulli nodes, and network coverage.

6.1 The CTR for k-connectivity

The k-connectivity graph property is an immediate extension of the concept of graph connectivity. Formally, k-connectivity is defined as follows (see also Appendix A):

Definition 6.1.1 (Connectivity) *A graph G is said to be k-connected, where $1 \leq k < n$, if for any pair of nodes u, v there exist at least k node disjoint paths connecting them. The connectivity of G, denoted as $\kappa(G)$, is the maximum value of k such that G is k-connected. A 1-connected graph is also called simply connected.*

A similar definition of connectivity can be given by considering edge, instead of node, disjoint paths between nodes. Denoting with $\xi(G)$ the edge-connectivity of G, it is seen immediately that $\kappa(G) \leq \xi(G)$. Figure 6.1 illustrates the concepts of k-connectivity and k-edge connectivity.

The interest in studying the CTR for k-connectivity is motivated by the fact that, when a network is k-connected, at most $k - 1$ node or link faults can be tolerated without disconnecting the network. So, a k-connected network is more resilient to faults than a simply connected network, where a single node or link failure might partition the network.

A network satisfying k-connectivity in general achieves also a better load balancing with respect to a simply connected network: in fact, messages between any two nodes u and v can be routed along at least k different paths, instead of along at least one single

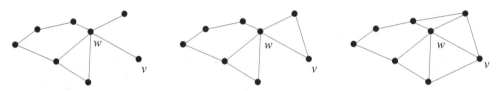

Figure 6.1 Simple and 2-connectivity. The graph on the left is simply connected (removing node w, or edge (w, v), is sufficient to disconnect the network). The graph in the center is 2-edge-connected, but not 2-(node)connected. In fact, removing any edge does not disconnect the graph, but removing node w does disconnect the graph. The graph on the right is 2-connected: removing any node or edge does not disconnect the graph.

path. In turn, better load balancing means a more evenly distributed energy consumption in the network, which potentially results in a longer network lifetime.

On the other hand, a connectivity value that is too high is detrimental for network capacity since any transmission would interfere with a large number of nodes. For instance, if $\kappa(G) = \frac{n}{2}$, it is seen immediately that any node in the communication graph has at least $\frac{n}{2}$ neighbors. In turn, this implies that when any node transmits, it interferes with at least $\frac{n}{2}$ nodes, and the network traffic carrying capacity is compromised. Thus, from a practical point of view, only networks with relatively low connectivity (say, below 5) are of some interest.

The first study of k-connectivity that can be applied to ad hoc networks is due to Penrose. In (Penrose 1999a), Penrose shows that the giant component phenomenon occurs in case of k-connectivity also, for any constant $1 \leq k < n$. More formally, Penrose proved the following theorem.

Theorem 6.1.2 (Penrose 1999a) *Assume n nodes are distributed uniformly at random in $R = [0, 1]^d$, with $d = 2, 3$. Let ρ_n (respectively, σ_n) denote the minimum value of the transmitting range at which the communication graph becomes k-connected (respectively, has minimum degree k), where $1 \leq k < n$ is an arbitrary constant. Then,*

$$\lim_{n \to \infty} P[\rho_n = \sigma_n] = 1.$$

In words, Theorem 6.1.2 states that, with high probability, the network becomes k-connected when the minimum node degree in the communication graph becomes k. Besides the important practical implications already discussed in Section 4.1, Theorem 6.1.2 proved useful in the characterization of the CTR for k-connectivity, which can be derived by analyzing the probability of the relatively simpler event that every node in the network has degree at least k. The value of the CTR for k-connectivity, which was partially characterized in (Penrose 1999a), has been recently derived in (Wan and Yi 2004) in case of two-dimensional networks.

Theorem 6.1.3 (Wan and Yi 2004) *Assume n nodes are distributed uniformly at random in the unit square $R = [0, 1]^2$. The CTR for k-connectivity, for any constant k, with $1 < k < n$, is*

$$r_k = \sqrt{\frac{\log n + (2k - 3) \log \log n + f(n)}{\pi n}},$$

where $f(n)$ is a function such that $\lim_{n \to \infty} f(n) = +\infty$.

Wan and Yi proved that a similar expression holds when nodes are uniformly distributed in the disk of unit area.

Comparing the expression of the CTR for k-connectivity with that of the CTR for simple connectivity (Corollary 4.1.2), we see that the difference between the two values is only in the second-order term $(2k - 3) \log \log n$ (we recall that k is a constant). This means that, asymptotically, k-connectivity with $k > 1$ is achieved by slightly increasing the transmitting range with respect to the critical value for simple connectivity.

The CTR for k-connectivity has also been studied under the assumption that n nodes are distributed in a two-dimensional region A with very large area (Bettstetter 2002). With this assumption, the number of nodes per units of area is $\rho = \frac{n}{a}$ with high probability, where a is the area of A. The following result has been proven in (Bettstetter 2002).

Theorem 6.1.4 (Bettstetter 2002) *Assume n nodes, each with transmitting range r_0, are distributed uniformly at random in A, where A has a very large area. The probability that the minimum node degree in the communication graph is at least k, for some $1 \leq k < n$, is closely approximated by*

$$P(deg_{\min} \geq k) \approx \left(1 - \sum_{i=0}^{k-1} \frac{(\rho \pi r_0^2)^i}{i!} \cdot e^{-\rho \pi r_0^2} \right)^n,$$

a.a.s., where $\rho = \frac{n}{a}$.

Given Theorem 6.1.2, the expression reported in Theorem 6.1.4 is also a close approximation of the probability of having a k-connected network.

Besides deriving the approximation of the probability of k-connectivity, the paper (Bettstetter 2002) also reports simulation results, which can be used to better understand the relative increase in the transmitting range needed to achieve k-connectivity, instead of simple connectivity. For instance, assuming that 500 nodes are uniformly distributed in a square of side 1000 m, setting the transmitting range to 90 m, corresponds to a probability of generating a simply connected graph equal to 0.9. In order to have the same probability of generating a 2-connected graph, the transmitting range must be set to approximately 107 m; for 3-connectivity, the transmitting range must be approximately 120 m. Thus, an approximately 19% increase with respect to the critical range for simple connectivity is sufficient to provide 2-connectivity, while an approximately 33% increase is sufficient for 3-connectivity. So, as predicted by Theorem 6.1.3, a relatively small increase of the transmitting range with respect to the critical value for connectivity is enough to achieve k-connectivity (for small values of $k > 1$).

6.2 The CTR for Connectivity with Bernoulli Nodes

The point graph model with *Bernoulli nodes* is an extension of the traditional point graph model. In this model, it is assumed that at any instant of time any node in the network is active with a certain constant probability $p > 0$. Since node activations are independent events, the node's active/inactive status can be modeled by a Bernoulli random variable of parameter p (this explains the name of the model).

Assume n nodes are distributed in a certain region R, each with transmitting range r and probability of being active equal to $p > 0$. We denote by $G(n, r)$ the communication graph

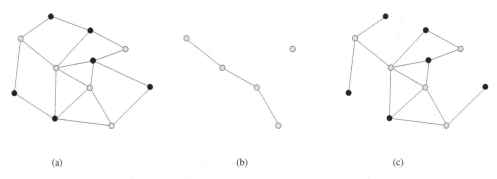

(a)   (b)   (c)

Figure 6.2 Example of $G(n, r)$ graph (a) and of its $A(n, r, p)$ (b) and $I(n, r, p)$ (c) subgraphs. Active nodes are light gray, and inactive nodes are black.

generated as in the traditional point graph model, that is, the graph obtained by connecting any two nodes that are at distance of, at most, r, independent of their active/inactive status. We denote the subgraph of $G(n, r)$ induced by the set of active nodes as $A(n, r, p)$. We denote as $I(n, r, p)$ the subgraph of $G(n, r)$ obtained from $G(n, r)$ by removing all links whose both endpoints are inactive nodes. An example of graph $G(n, r)$, and of its subgraphs $A(n, r, p)$ and $I(n, r, p)$, is reported in Figure 6.2.

Recent papers have investigated asymptotic conditions under which $A(n, r, p)$ and $I(n, r, p)$ are connected with high probability. The motivation for analyzing the connectivity of these graphs stems from the fact that $A(n, r, p)$ and $I(n, r, p)$ can be used to model several network design problems, such as the following:

– *Randomized virtual backbone construction*: In many applications of WSNs, nodes alternately shut down their transceivers in order to reduce power consumption. (We recall that the power consumption of a sensor node can be considerably reduced by turning the radio off–see Section 2.3). However, a certain number of nodes must keep the radio on, in order to preserve network connectivity. Thus, active nodes must form a connected backbone. We refer to this property as 'active connectivity'. Another desirable property is that any inactive node has at least one active node within its transmitting range. In fact, inactive nodes still sense the environment (it is only the radio apparatus that is turned off), and, in case an inactive node detects an anomalous event, we want that the information regarding this event propagates quickly through the network, eventually reaching the operator. This can be accomplished only if every inactive node is able to directly communicate with at least one active node (and if the set of active nodes forms a connected backbone). Since if this property holds the set of active nodes is a dominating set, we refer to this property as 'active domination'. Examples of virtual backbones are reported in Figure 6.3.

A simple randomized strategy to build a virtual backbone of active nodes is as follows: any node in the network remains active for a fraction $0 < p \leq 1$ of its operational time, where the activation periods are randomly chosen. Assume that n nodes are distributed in a certain region R, and each node has the same transmitting range r. It is seen immediately that the virtual backbone resulting from the randomized strategy above satisfies active connectivity if and only if graph $A(n, r, p)$ is connected, and

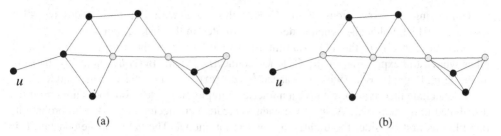

(a)   (b)

Figure 6.3 Active connectivity and active domination of the virtual backbone. Active nodes are light gray, and inactive nodes are black. The backbone of active nodes in (a) satisfies active connectivity, but not active domination (node u has no direct connection to any active node). The backbone in (b) satisfies both active connectivity and active domination.

that it satisfies both active connectivity and active domination if and only if graph $I(n, r, p)$ is connected.

- *Randomized broadcast*: Assume a certain network node u wants to broadcast a message m. Performing broadcast in ad hoc networks is a nontrivial task, because of the problem of spatial reuse: if many nodes try to relay m simultaneously, it is likely that they corrupt each other's transmission, leading to an increase in the broadcasting latency and/or energy consumption. This problem is known in the literature as the *broadcast storm* problem (see Chapter 8 for a more detailed description of this phenomenon). An easy strategy to prevent the broadcast storm problem is to use randomization: when a node receives message m, it relays m with a certain probability $0 < p \leq 1$, independent of every other node. It is easy to see that under the assumption that n nodes with transmitting range r are distributed in a certain region message m eventually reaches all the network nodes if and only if graph $I(n, r, p)$ is connected.

The connectivity of graphs $A(n, r, p)$ and $I(n, r, p)$ can be characterized by combining Theorem 9 of (Yi et al. 2003) and Theorem 9 of (Yi and Wan 2005).

Theorem 6.2.1 *Assume n nodes are distributed uniformly at random in the disk of unit area. Let $r_n(\xi) = \sqrt{\frac{\log n + \xi}{\pi p n}}$, for some constant ξ, and let ρ_A (respectively, ρ_I) be the minimum transmitting range such that graph $A(n, \rho_A, p)$ (respectively, $I(n, \rho_I, p)$) is connected. Then,*

$$\lim_{n \to \infty} P(\rho_A \leq r_n(\xi)) = \exp(-pe^{(-\xi)}),$$

$$\lim_{n \to \infty} P(\rho_I \leq r_n(\xi)) = \exp(-e^{(-\xi)}).$$

Corollary 6.2.2 *Assume n nodes are distributed uniformly at random in the disk of unit area, and assume that nodes are active with constant, independent probability p, with $0 < p \leq 1$. The CTR for connectivity of $A(n, r_n, p)$ and of $I(n, r_n, p)$ is the same and equals*

$$r_{BN} = \sqrt{\frac{\log n + f(n)}{\pi p n}},$$

where $f(n)$ is an arbitrary function such that $\lim_{n \to \infty} f(n) = +\infty$.

Comparing the expressions of the CTR without and with Bernoulli nodes (corollaries 4.1.2 – which holds also when nodes are distributed in the disk of unit area – and 6.2.2), the only difference is in the additional multiplicative term p at the denominator of r_{BN}. In other words, the expression of the CTR for connectivity with Bernoulli nodes is the same as in the traditional model, with n replaced by pn (expected number of active nodes).

To conclude this section, we give a numeric example. Suppose 1000 nodes are uniformly distributed in the unit disk. Assume we want to create a connected network with probability 0.99. Let us first consider the traditional point graph model. The value of the constant β in Theorem 4.1.1 such that $\exp(-e^{-\beta}) = 0.99$ is approximately 4.6. With this value of β, we get a value of the transmitting range equal to 0.060523. Assume now that nodes are active with probability $p = 0.5$. In order to have probability 0.99 that $A(1000, r, 0.5)$ is connected, we must set r to 0.0829867, which is an approximately 37% increase with respect to the case of always active nodes. In order to have the same probability that $I(1000, r, 0.5)$ is connected, we must set r to 0.0855924, which is an approximately 41% increase with respect to the case of always active nodes.

6.3   The Critical Coverage Range

The Critical Coverage Range (CCR) problem is defined as follows:

Definition 6.3.1 (Critical coverage range) *Assume n nodes are deployed into a certain region R. A point x in region R is said to be covered if it is at a distance of, at most, r from at least one of the network nodes, where r is the nodes' covering range. We say that region R is covered if all of its points are covered. The CCR problem is to find, given a node deployment, the minimum value of r such that R is covered.*

Similar to the CTR problem, the CCR problem can be easily solved if nodes' positions are known. Furthermore, it can be formulated also in the reverse way, that is: assume a certain region R must be covered using nodes with sensing range r; which is the minimum number n of nodes to be deployed in order to cover R?

The study of the CCR problems stated above finds its motivation in the context of wireless sensor networks used for monitoring applications, such as surveillance or habitat monitoring. In the design of this type of networks, it is often assumed that every node (sensor) can 'sense' an event within a certain maximum range (the coverage range), and the typical requirement is that the monitored region is covered. Since sensor nodes in this context are typically randomly deployed (for instance, using a moving vehicle such as airplane), the CCR is studied under the assumption of random node deployment.

The reader would have noticed the strong similarities between the CTR and the CCR problem. Indeed, it is easy to prove that a node deployment that covers R under the assumption that nodes have coverage range r_c also generates a connected communication graph under the assumption that nodes have transmitting range $r_t \geq 2r_c$ (see Figure 6.4). This is formally stated in the theorem below, which has been proven in (Wang et al. 2003).

Theorem 6.3.2 (Wang et al. 2003) *Assume that a set S of n nodes with coverage range r_c and transmitting range $r_t \geq 2r_t$ are deployed in a certain region R and that the nodes in S cover R. Then, the communication graph generated by nodes in S is connected.*

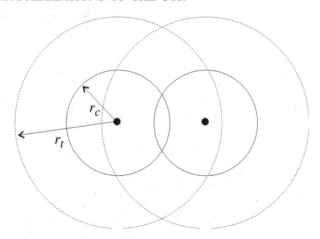

Figure 6.4 Relation between the coverage range (r_c) and the transmitting range (r_t): setting $r_t = 2r_c$, the covering ranges of two nodes overlap if and only if they are in each other transmitting range.

Note that the reverse of the theorem above does not hold. This is depicted in Figure 6.5: the communication graph formed by the nodes in S is connected, but the region R is not covered. This example shows that coverage is, in general, a stronger requirement than connectivity, even when $r_t \geq 2r_c$: a set of nodes that is concentrated in a subregion of R can be connected, but it does not satisfy coverage (Figure 6.5).

The critical coverage range has been investigated in (Philips et al. 1989) for the case of nodes distributed in a square with side of length l according to a Poisson process of fixed density λ.

Theorem 6.3.3 (Philips et al. 1989) *Assume nodes are distributed in $R = [0, l]^2$ according to a two-dimensional Poisson process of density $\lambda > 0$. Let r_c denote the coverage range of the nodes. If*

$$r_c = \sqrt{\frac{2(1 - \varepsilon) \log l}{\pi \lambda}},$$

for some $0 < \varepsilon < 1$, then R is a.a.s. not covered (i.e. $\lim_{l \to \infty} P[R$ is covered$] = 0$). If

$$r_c = \sqrt{\frac{2(1 + \varepsilon) \log l}{\pi \lambda}},$$

for some $0 < \varepsilon < 1$, then R is a.a.s. covered (i.e. $\lim_{l \to \infty} P[R$ is covered$] = 1$).

Note that, with respect to the characterization of the CTR for uniformly distributed nodes (Corollary 4.1.2), the result stated in Theorem 6.3.3 is somewhat weaker: instead of an additive term (function $f(n)$ in the statement of Corollary 4.1.2), we have a multiplicative constant c. If $c < 2$, then R is not covered a.a.s., while if $c > 2$ a.a.s. coverage holds. However, whether R is covered when $c = 2$ is an open question.

A more direct relation between the CTR and the CCR for Poisson distributed points has been derived for one-dimensional networks. The theorem below is due to Piret (Piret 1991).

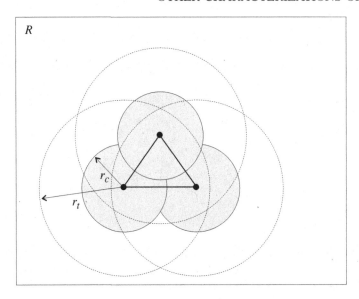

Figure 6.5 Example of node deployment which generates a connected network (bold edges), but does not satisfy coverage: only the shaded subregion of R is covered by at least one node.

Theorem 6.3.4 (Piret 1991) *Assume nodes are distributed in $R = [0, l]$ according to a one-dimensional Poisson process of density $\lambda > 0$. Let r_c (respectively, r_t) denote the coverage range (respectively, the transmitting range) of the nodes. If*

$$r_c = \frac{(1 - \varepsilon) \log l\lambda}{2\lambda},$$

for some $0 < \varepsilon < 1$, then R is a.a.s. not covered. If

$$r_c = \frac{(1 + \varepsilon) \log l\lambda}{2\lambda},$$

for some $0 < \varepsilon < 1$, then R is a.a.s. covered. If

$$r_t = \frac{(2 - \varepsilon) \log l\lambda}{2\lambda},$$

for some $0 < \varepsilon < 1$, then the resulting communication graph is a.a.s. disconnected. If

$$r_c = \frac{(2 + \varepsilon) \log l\lambda}{2\lambda},$$

for some $0 < \varepsilon < 1$, then the resulting communication graph is a.a.s. covered.

Theorem 6.3.4 is very important, since it states that, at least in the case of Poisson distributed points on a line, the CCR is equivalent to the CTR problem with r_t replaced by $2r_c$. In other words, under these assumptions the probability that a subregion of R remains uncovered when the network is a.a.s. connected is asymptotically negligible. Whether the same holds for two-dimensional networks, or with different node distributions, is an open problem.

Part III

Topology Optimization Problems

7

The Range Assignment Problem

In Chapters 4, 5, and 6, we have investigated various network design problems under the assumption that all the nodes have the same transmitting range, which reflects all those situations in which nodes cannot change the transmit power level (and use transceivers with the same technology). However, in many scenarios, nodes *can* change the transmit power level. So, the problem of choosing the nodes' transmit power levels in such a way that the network topology satisfies certain properties becomes relevant. In this chapter, we consider the problem of determining a set of power level assignments that generates a connected communication graph while at the same time minimizing the energy consumption. This problem is known in the literature as the *Range Assignment problem*.

7.1   Problem Definition

We recall that, given the set N of network nodes, a range assignment for N is a function RA that assigns to every $u \in N$ a transmitting range $RA(u)$, with $0 < RA(u) \leq r_{max}$, where r_{max} is the maximum transmitting range. Note that, under the assumption that the path loss model is the same for all the network nodes, and that shadowing/fading effects are not considered, transmitting range, and transmit power level are equivalent concepts. Since traditionally the function RA is defined in terms of range, instead of power, we keep this convention.

The Range Assignment problem, which was first studied in (Kirousis et al. 2000), is defined as follows:

Definition 7.1.1 (RA problem) *Let N be a set of nodes in the d-dimensional space, with $d = 1, 2, 3$. Determine a range assignment function $\overline{RA}$ such that the corresponding communication graph is strongly connected, and $c(\overline{RA}) = \sum_{u \in N}(\overline{RA}(u))^{\alpha}$ is minimum over all connecting range assignment functions, where α is the distance-power gradient.*

The cost measure $c(RA)$ used in the definition of the RA problem is the sum of the transmit power levels used by all the nodes in the network. Thus, RA can be informally stated as the problem of finding a 'minimal' nodes' range assignment that generates a connected communication graph, where 'minimal' is intended as 'least energy cost'. Besides reducing energy consumption, a connecting range assignment with minimum energy cost is likely to increase network capacity also, for the reasons discussed in Chapter 3. These observations motivate the interest in studying the RA problem.

In a certain sense, the RA problem can be seen as a generalization of the problem of determining the CTR for connectivity, where the constraint that all the nodes have the same transmitting range is dropped. As we shall see, dropping this constraint considerably increases the complexity of finding the optimal solution.

7.2   The RA Problem in One-dimensional Networks

The optimal solution to the RA problem can be found in polynomial time in case of one-dimensional networks. In this Section, we present the algorithm for finding the optimal solution introduced in (Kirousis et al. 2000).

Before presenting the algorithm, we need some preliminary definitions.

Let $N = \{u_1, \ldots, u_n\}$ be a set of colinear points (nodes). Without loss of generality, assume that nodes are increasingly ordered according to their spatial coordinate, that is, u_1 is the leftmost node and u_n is the rightmost node. Given a set of nodes and a range assignment RA, we say that edge (u_i, u_j) in the resulting communication graph is *backward* if $i > j$, that is, if the edge goes from right to left. For any i, j with $1 \leq i < j \leq n$, we define set $E_{i,j}$ as the set of all the backward edges that have both their endpoints in $\{u_i, \ldots, u_j\}$. Formally, $E_{i,j} = \{(u_s, u_r) : i \leq r < s \leq j\}$. An example of backward edge set is reported in Figure 7.1.

The algorithm for finding the optimal solution is based on a recursive construction: given the optimal connecting range assignment for nodes $\{u_1, \ldots, u_k\}$, for some $1 \leq k < n$, a strategy is given to build the optimal assignment for the set of nodes $\{u_1, \ldots, u_{k+1}\}$.

The intuition behind the recursive strategy is the following: when the optimal solution RA_k at step k is given and the solution for the next step is to be determined, the cost of RA_k can be considered as zero (by hypothesis, at least cost $c(RA_k)$ is necessary to connect the k leftmost nodes), and the 'minimal increase' to the range assignment that connects node u_{k+1} also must be identified. This leads to the following definition of incremental cost of a range assignment:

Definition 7.2.1 (RA incremental cost) *Let* $N = \{u_1, \ldots, u_n\}$ *be a set of nodes, and* E *a set of directed edges between nodes in* N. *The range assignment induced by set* E, *denoted by* RA_E, *is the minimal assignment such that* $RA_E(u_i) \geq \delta(u_i, u_j)$, *for any directed edge* $(u_i, u_j) \in E$. *The incremental cost of range assignment* RA *with respect to* E, *denoted by* $c_E(RA)$ *is defined as* $c_E(RA) = \sum_{i:RA_E(u_i) \neq RA(u_i)} (RA(u_i))^\alpha$. *We say that edges in* E *are free of cost with respect to range assignment* RA.

Figure 7.1   Backward edges in the set $E_{2,5}$ (bold edges).

The strategy is based on the following recursive assumption: for any $j \leq k$ and any $l \geq k$, there exists a range assignment RA_k with minimum cost among those that generate a communication graph with the following properties:

1. There is a path between any pair of nodes in $\{u_1, \ldots, u_k\}$.

2. There exists directed edge (u_i, u_l), for some $1 \leq i \leq k$.

3. Any (backward) edge in $E_{j,k}$ is free of cost with respect to RA_k.

Let (N', E) be a directed graph, where $N' \subset N$, and let v be an additional node in N, which we call the *receiver* node. A range assignment RA is said to be *total* for $((N', E), v)$ if and only if

1. the graph on node set N' obtained by adding to E the edges induced by range assignment RA (i.e. edges (u_i, u_j) such that $RA(u_i) \geq \delta(u_i, u_j)$) is strongly connected;

2. there exists directed edge (u_i, v), for some $u_i \in N'$; that is, $RA(u_i) \geq \delta(u_i, v)$ for some $u_i \in N'$.

The cost of a total range assignment for $((N', E), v)$ is the incremental cost with respect to RA_E, that is, $c_E(RA)$. Intuitively, a total range assignment has zero cost for the edges in E, and establishes communication paths between any pair of nodes in N', and also between a node in N' and the receiver, in this direction only. A total range assignment for $((N', E), v)$ of minimum cost is said to be *optimal*. In the following, $Feas((N', E), v)$ denotes the set of total range assignments for $((N', E), v)$, and $Opt((N', E), v)$ denotes the set of optimal range assignments for $((N', E), v)$. Finally, given $u \in N'$ and a positive real r, we denote with $Opt((N', E), v, (u, r))$ the set of range assignments of minimum cost among the assignments $RA \in Feas((N', E), v)$ such that $RA(u) = r$.

The OPTIMAL1DRA algorithm for finding the optimal solution to the RA problem in one-dimensional networks is reported in Figure 7.2. The algorithm first identifies a set of optimal range assignments for connecting node u_1 to any other single node (step 1.2). Then, we have the recursive step, in which a set of optimal range assignments for connecting nodes in $\{u_1, \ldots, u_k\}$ with a receiver node u_l, with $k \leq l \leq n$, is calculated. For details on how optimal range assignments are calculated (step 2.4), the reader is referred to Lemma 2.6 of (Kirousis et al. 2000). After n recursive steps, any range assignment in $Opt(((\{u_1, \ldots, u_n\}, \emptyset), u_n)$ is optimal for N.

The correctness of OPTIMAL1DRA has been proven in (Kirousis et al. 2000). The authors also proved that the computational complexity of the algorithm is $O(n^4)$.

Comparing the computational complexity of OPTIMAL1DRA to that of an algorithm for finding the critical range for connectivity, we can observe the increase in complexity caused by dropping the assumption that all the nodes have the same transmitting range. In case of colinear points, the CTR can be found in $O(n \log n)$ time,[1] which should be compared to the $O(n^4)$ running time of OPTIMAL1DRA. The gap in terms of computational complexity is considerable: when $n = 100$, the running time of the optimal algorithm increases from about 1000 time units in case of the CTR problem to about 10^8 time units in case of the one-dimensional RA problem.

---

[1] An algorithm for finding the CTR in $O(n \log n)$ time in one-dimensional networks is the following. First, order all the nodes according to their spatial coordinate. The CTR for connectivity is the largest among the distances between consecutive nodes in the order.

Algorithm OPTIMAL1DRA:

1. Initialization

1.1   Let RA_i be the range assignment such that $RA_i(u_1) = \delta(u_1, u_i)$,
       and $RA_i(u_j) = 0$ otherwise

1.2   for $i = 2, \ldots, n$ do $Opt((\{u_1\}, \emptyset), u_i) = RA_i$

2. Step k:

2.1   Assume we know $Opt((\{u_1, \ldots, u_k\}, E_{i,k}), u_l)$, for any $1 \leq i \leq k$ and $k \leq l \leq n$

2.2   for any j, m such that $1 \leq j \leq k + 1 \leq m \leq n$

2.3   consider all possible values of $RA(u_{k+1})$ (there are $k + 2$ such values)

2.4   for each such value r, find an assignment $\overline{RA}$ in
           $Opt((\{u_1, \ldots, u_k\}, E_{j,k+1}), u_{k+1}, (u_{k+1}, r))$

2.5   if $\overline{RA}$ has cost lower than that of the current range assignment for j, m,
           store $\overline{RA}$ (new current minimum)

2.6   at the end of step k, we know a range assignment in
       $Opt((\{u_1, \ldots, u_{k+1}\}, E_{i,k+1}), u_l)$, for any $1 \leq i \leq k + 1 \leq l \leq n$

3. after step n, an optimal assignment is one in $Opt((\{u_1, \ldots, u_n\}, \emptyset), u_n)$

Figure 7.2 Algorithm for finding the optimal range assignment in one-dimensional networks.

7.3   The RA Problem in Two- and Three-dimensional Networks

In the previous section, we have analyzed the RA problem for one-dimensional networks, outlining the considerable increase in computational complexity with respect to the case of solving the simpler CTR problem. The increase in computational complexity becomes even larger in case of two- and three-dimensional networks, as stated by the following theorem.

Theorem 7.3.1 *Solving the RA problem in two- and three-dimensional networks is NP-hard.*

The NP-hardness of finding the optimal solution to RA in three-dimensional networks has been proved in (Kirousis et al. 2000). Later on, Clementi et al. proved that the problem remains NP-hard in case of two-dimensional networks also (Clementi et al. 1999).

Although solving RA in two- and three-dimensional networks is hard, an approximation of the optimal solution can be easily computed by constructing an MST on the nodes. The construction of the range assignment is as follows:

Let $N = \{u_1, \ldots, u_n\}$ be a set of points (nodes) in the two- or three-dimensional space.

1. Construct an undirected weighted complete graph $G = (N, E)$, where the weight of edge $(u_i, u_j) \in E$ is $\delta(u_i, u_j)^\alpha$.

2. Find a minimum weight spanning tree T of G.

3. Define range assignment RA_T, with $RA_T(u_i) = \max_{j|(u_i,u_j)\in T} \delta(u_i, u_j)$.

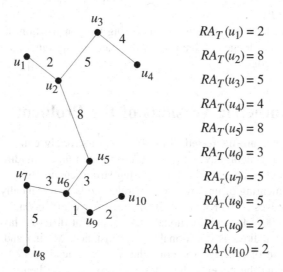

$$RA_T(u_1) = 2$$

$$RA_T(u_2) = 8$$

$$RA_T(u_3) = 5$$

$$RA_T(u_4) = 4$$

$$RA_T(u_5) = 8$$

$$RA_T(u_6) = 3$$

$$RA_T(u_7) = 5$$

$$RA_T(u_8) = 5$$

$$RA_T(u_9) = 2$$

$$RA_T(u_{10}) = 2$$

Figure 7.3 Minimum spanning tree T on the set of nodes, and corresponding range assignment RA_T.

An example of minimum spanning tree T, and the corresponding range assignment RA_T, are depicted in Figure 7.3.

The algorithm for constructing RA_T has $O(n^2)$ running time (the time complexity of building the MST on the n nodes), and produces a 2-approximation of the optimal solution.

Theorem 7.3.2 (Kirousis et al. 2000) *Let N be a set of points (nodes) in the two- or three-dimensional space, and let RA_T be the range assignment defined as above. Let $\overline{RA}$ be an optimal range assignment for the RA problem. Then*

$$c(RA_T) < 2c(\overline{RA}).$$

Proof. The proof is composed of two steps. First, we prove that $c(\overline{RA})$ is greater than the cost $c(T)$ of the minimum spanning tree T. Then, we prove that $c(RA_T) < 2c(T)$.

1. $c(\overline{RA}) > c(T)$

Starting from any optimal assignment $\overline{RA}$ for N, we can build a spanning tree for the complete undirected graph G by choosing any node $u \in N$, and constructing a shortest path destination tree rooted at u, with all edges directed toward the root, representing minimum weight paths from any node to the root node. Given the shortest path destination tree, the corresponding spanning tree T' is obtained by changing the directed edges in the shortest path tree to the corresponding undirected edges in G. Since each of the $n - 1$ nodes other than the root must be assigned a range that is at least sufficient to establish the edges in the shortest path destination tree, we have $c(\overline{RA}) > c(T')$. The strict inequality follows from the fact that $\overline{RA}$ assigns a strictly positive range to the root node u (this is necessary for strong connectivity), which is not accounted for in $c(T')$. In turn, $c(T')$ is at least as large as the cost $c(T)$ of the minimum spanning tree.

2. $c(RA_T) < 2c(T)$

The inequality follows by observing that, during the construction of RA_T, each edge of T can be chosen as the 'longest' edge (i.e. as the transmitting range) at most by two nodes (the endpoints of the edge).

7.4 The Symmetric Versions of the Problem

In the RA problem we are interested in establishing a strongly connected communication graph. Since nodes in general have different transmitting ranges, unidirectional links might occur, and they can even be essential for ensuring strong connectivity.

Although implementing of unidirectional wireless links is technically feasible (see (Bao and Garcia-Luna-Aceves 2001; Kim et al. 2001; Pearlman et al. 2000a; Prakash 2001; Ramasubramanian et al. 2002) for unidirectional link support at different layers), the advantage of using unidirectional links is questionable. For instance, Marina and Das have recently observed that, in case of routing protocols, the high overhead needed to handle unidirectional links outweights the benefits that they can provide, and better performance can be achieved by simply avoiding them (Marina and Das 2002).

Indeed, most routing protocols for ad hoc networks (for instance, DSR (Johnson et al. 2002) and AODV (Perkins et al. 2002)) are based on the implicit assumption that wireless links can be 'reversed', that is, must be bidirectional. The same observation applies to the current implementation of the MAC layer in the IEEE 802.11 standard, which is based on a RequestToSend/ClearToSend message exchange: when node u wishes to send a message to a node v within its transmitting range, it sends a RTS to v and waits for the CTS message from v. If the CTS is not received within a certain period of time, the message transmission is aborted and it is tried again after a backoff interval. If the wireless link between nodes u and v is unidirectional, either one of the RTS or CTS message is not received, and communication is not possible. Supporting unidirectional links at the MAC layer would imply that intermediate nodes should relay the RTS/CTS messages on behalf of node u or v. Alternatively, a different channel access mechanism (for instance, based on collision detection instead of collision avoidance) should be used. Anyway, supporting unidirectional links would imply a considerable modification of the current implementation of the IEEE 802.11 MAC protocol.

The reasons above have motivated researchers to investigate restricted versions of the RA problem, where certain symmetry constraints are imposed on the communication graph. In particular, the following two problems have been defined and investigated (Blough et al. 2002; Calinescu et al. 2002):

Definition 7.4.1 (WSRA problem) *Let N be a set of nodes in the d-dimensional space, with $d = 1, 2, 3$. Let RA be a range assignment for N and let G be the corresponding (directed) communication graph. The symmetric subgraph of G, denoted by G_S, is the undirected graph obtained from G by removing unidirectional links. The WSRA problem is to determine a range assignment function $\overline{RA}$ such that G_S is connected, and $c(\overline{RA}) = \sum_{u \in N} (\overline{RA}(u))^\alpha$ is minimum, where α is the distance-power gradient.*

Definition 7.4.2 (SRA problem) *Let N be a set of nodes in the d-dimensional space, with $d = 1, 2, 3$. A range assignment RA for N is said to be symmetric if it generates a*

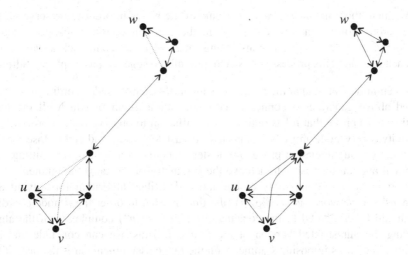

Figure 7.4 The different symmetry requirements in the WSRA and in the SRA problem. In WSRA, unidirectional links (dashed edges) are allowed, but they are not essential for connectivity. In SRA, all the links in the communication graph must be bidirectional: nodes u, v, and w must increase their transmitting range to meet this stronger symmetry requirement.

communication graph that contains only bidirectional links, that is, $RA(u_i) \geq \delta(u_i, u_j) \Leftrightarrow RA(u_j) \geq \delta(u_i, u_j)$. The Symmetric Range Assignment (SRA) problem is to determine a SRA function $\overline{RA}$ such that the corresponding communication graph is connected, and $c(\overline{RA}) = \sum_{u \in N} (\overline{RA}(u))^\alpha$ is minimum, where α is the distance-power gradient.

Note the different symmetry requirements in the two versions of the problem: in the WSRA (Weakly Symmetric Range Assignment) problem, the communication graph may contain unidirectional links which, however, are not essential for connectivity. On the other hand, in the SRA problem, the communication graph must contain *only* bidirectional links. This is a much stronger requirement on the communication graph, as the example reported in Figure 7.4 shows. The motivation for studying WSRA stems from the observation that what is really important in the design of ad hoc and sensor networks is the existence of a connected backbone of symmetric edges. In other words, there could exist links for which symmetry is not guaranteed, but these links can be ignored without compromising network connectivity.

7.4.1 The SRA problem in one-dimensional networks

In case of colinear nodes, the optimal SRA for a set of nodes can be constructed as follows:

1. Order the nodes according to their spatial coordinate; let $\{u_1, \ldots, u_n\}$ be the resulting node ordering.

2. Assign to node u_1 transmitting range $\delta(u_1, u_2)$, to node u_n transmitting range $\delta(u_{n-1}, u_n)$, and to every other node u_i transmitting range equal to $\max\{\delta(u_{i-1}, u_i), \delta(u_i, u_{i+1})\}$.

3. Augment the transmitting range of some of the nodes in order to preserve symmetry: for any unidirectional edge (u_i, u_j) in the communication graph generated at the previous step, increase the transmitting range of node u_j in such a way that it can reach node u_i. This process is repeated until all the edges in the graph are bidirectional.

It is seen immediately that the range assignment $\overline{RA}$ constructed according to the strategy described above generates a connected communication graph in which all the links are bidirectional. To prove that $\overline{RA}$ is optimal, it is sufficient to observe that, in order to achieve connectivity, every node must be connected at least to its left and right closest neighbor; furthermore, the augmentation procedure at step 3 increases a node's transmitting range of the minimal amount necessary to achieve the symmetry of the range assignment.

The computational complexity of the above described algorithm for solving SRA in one-dimensional networks is $O(n \log n)$ (the time needed to order the n node coordinates), which should be compared to the considerably higher $O(n^4)$ complexity of the algorithm for solving the unrestricted version of the problem. Thus, we can conclude that in one-dimensional networks imposing symmetry on the range assignment eases the task of finding the optimal solution.

7.4.2   The SRA problem in two- and three-dimensional networks

In this section, we show that, contrary to the case of one-dimensional networks, in two-, and three-dimensional networks imposing the symmetry condition on the range assignment does not change the computational complexity of the problem. As the reader will notice, the proof (presented in (Blough et al. 2002)) is quite lengthy and complicated. The difficulty of the proof stems from the fact that, when studying the complexity of ad hoc network problems, *geometry cannot be ignored*. In other words, when considering reductions from known NP-hard problems (the MINWEIGHTEDVERTEXCOVER problem in the example below) to the problem at hand, we have to prove that nodes can actually be placed in the two- or three-dimensional space in such a way that *any instance* of the problem to be reduced can be transformed into a corresponding instance of the problem at hand. This is usually accomplished by making use of a geometric construction, or *gadget*.

For ease of presentation, assume $\alpha = 2$. In order to prove the NP-hardness of SRA, we will show a polynomial-time reduction from MINWEIGHTEDVERTEXCOVER for planar cubic graphs, which is known to be NP-hard (Garey and Johnson 1977). The proof is based on a modification of the construction used in (Clementi et al. 1999) to prove that solving RA in two-dimensional networks is NP-hard. The construction can be summarized as follows:

- Given a planar cubic graph[2] G, construct a planar orthogonal drawing of G.

- Add two new vertices for each bend of the drawing so to obtain a straight-line drawing $D(G)$.

- Replace each straight-line (edge) in $D(G)$ with a suitable set of nodes (gadget). The set of points in the two-dimensional space resulting from this replacement is denoted by $S(G)$.

---

[2] A graph is cubic if every node in it has degree three. A graph is planar if it can be drawn in the plane in such a way that no two edges cross each other.

The following properties characterize gadgets (Clementi et al. 1999):

Let $D(G) = (V, E)$ be a straight-line planar orthogonal drawing of a planar cubic graph G. Let $\lambda, \lambda', \varepsilon \geq 0$ be such that $\lambda + \varepsilon > \lambda'$, and let $\gamma > 1$. For any $(a, b) \in E$, the corresponding gadget g_{ab} is formed by the disjoint sets of points $V_{ab} = \{a, b\}$, $Y_{ab} = \{y_{ab}, y_{ba}\}$, $X_{ab} = \{x_1, \ldots, x_{l_1}\}$, and $Z_{ab} = \{z_1, \ldots, z_{l_2}\}$, where l_1 and l_2 depend on the length of (a, b) in the drawing. These sets of points are drawn in $\mathbf{R}^2$ so that the following properties hold:

(a) $\delta(a, y_{ab}) = \delta(b, y_{ba}) = \lambda + \varepsilon$.

(b) X_{ab} is a chain of points drawn so that $\delta(a, x_1) = \delta(x_1, x_2) = \cdots = \delta(x_{l_1}, b) = \lambda$ and, for any $i \neq j + 1, j - 1, \delta(x_i, x_j) \geq \lambda$.

(c) Z_{ab} is a chain of points drawn so that $\delta(y_{ab}, z_1) = \delta(z_1, z_2) = \cdots = \delta(z_{l_2}, y_{ba}) = \lambda'$ and, for any $i \neq j + 1, j - 1, \delta(z_i, z_j) \geq \lambda'$.

(d) For any $x_i \in X_{ab}$, $z_j \in Z_{ab}$, $\delta(x_i, z_j) > \lambda + \varepsilon$. Furthermore, for any $i = 1, \ldots, l_1$, $\delta(x_i, y_{ab}) \geq \lambda + \varepsilon$ and $\delta(x_i, y_{ba}) \geq \lambda + \varepsilon$.

(e) Given any two different gadgets g_{ab} and g_{cd}, for any $v \in g_{ab} \setminus g_{cd}$ and $w \in g_{cd} \setminus g_{ab}$, we have that $\delta(v, w) \geq \lambda$. Furthermore, if $v \notin V_{ab} \cup X_{ab}$ or $w \notin V_{cd} \cup X_{cd}$ then $\delta(v, w) \geq \gamma \lambda$.

In (Clementi et al. 1999) it is shown that, for a suitable choice of constants $\lambda, \lambda', \varepsilon$ and γ, gadgets whose points have properties $(a) \ldots (e)$ can be drawn in $\mathbf{R}^2$ for any straight-line planar orthogonal drawing $D(G)$. It can be seen that the same choice of $\lambda, \lambda', \varepsilon$ and γ, achieves points in the gadgets to have the following additional properties:

(b') $\delta(a, x_j) > \lambda + \varepsilon$ for any $j \neq 1$, and $\delta(x_i, b) > \lambda + \varepsilon$ for any $i \neq l_1$.

(c') $\delta(y_{ab}, z_j) > \lambda + \varepsilon$ for any $j \neq 1$, and $\delta(z_i, y_{ba}) > \lambda + \varepsilon$ for any $i \neq l_2$.

(d') For any $x_i \in X_{ab}$, $z_j \in Z_{ab}$, $\delta(x_i, y_{ab}) > \lambda + \varepsilon$, $\delta(x_i, y_{ba}) > \lambda + \varepsilon$, $\delta(z_j, a) > \lambda + \varepsilon$, and $\delta(z_j, b) > \lambda + \varepsilon$.

Given properties $(a) \ldots (e)$ and $(b') \ldots (d')$, it turns out that every gadget consists of two components whose relative distance is $\lambda + \varepsilon$: the VX-component, consisting of the chain of points in $V_{ab} \cup X_{ab}$, and the YZ-component, consisting of the chain of points in $Y_{ab} \cup Z_{ab}$. Furthermore, given any pair of nodes (v, w) such that v is in the VX-component and w is in the YZ-component, we have that $\delta(v, w) = \lambda + \varepsilon$ if and only if $v = a$ and $w = y_{ab}$ or $v = b$ and $w = y_{ba}$. The gadget for edge (a, b) is depicted in Figure 7.5.

Observe that in a connecting range assignment any node must have a transmitting range at least equal to the distance to its closest neighbor. Let $RA_{\min}$ be the range assignment for $S(G)$ such that every node have a transmitting range equal to the distance to its closest neighbor. Given the properties of gadgets, $RA_{\min}$ is such that nodes in the VX-components have transmitting range λ, and nodes in the YZ-components have transmitting range λ'. Because of the symmetry of points in the plane, $RA_{\min}$ is symmetric. The communication graph induced by $RA_{\min}$ is composed of $m + 1$ connected components, where $m = |E|$: the YZ-components of the m gadgets and the union VX of all the VX-components of the gadgets. Hence, in order to have a connected and symmetric communication graph, we need to define some *bridge points* between VX and every YZ-component.

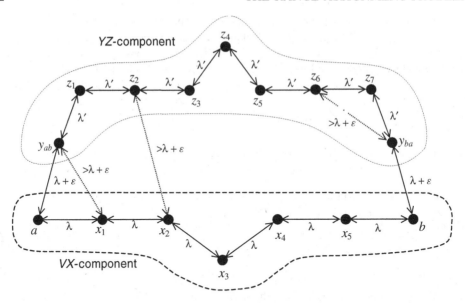

Figure 7.5 The gadget for edge (a, b).

Let $Y = \bigcup_{a,b \in E} Y_{ab}$, $X = \bigcup_{a,b \in E} X_{ab}$, $Z = \bigcup_{a,b \in E} Z_{ab}$, and $V = \bigcup_{a,b \in E} V_{ab}$. The following lemma characterizes the properties of the optimal symmetric range assignment for $S(G)$.

Definition 7.4.3 (Canonical RA) *A symmetric connecting range assignment RA_c for $S(G)$ is said to be canonical if*

- $RA_c(v) = \lambda$ *for any* $v \in X$;

- $RA_c(v) = \lambda'$ *for any* $v \in Z$;

- $RA_c(v) = \lambda$ *or* $RA_c(v) = \lambda + \varepsilon$ *for any* $v \in V$;

- $RA_c(v) = \lambda'$ *or* $RA_c(v) = \lambda + \varepsilon$ *for any* $v \in Y$.

Lemma 7.4.4 (Blough et al. 2002) *Let $S(G)$ be a set of points placed in $\mathbf{R}^2$ according to the above described construction, where γ, λ, and ε are positive constants such that*

$$(\gamma \lambda)^2 > \frac{m-1}{m} \left((\lambda + \varepsilon)^2 - \lambda^2 \right) + (\lambda + \varepsilon)^2 . \tag{7.1}$$

Then, for any symmetric connecting range assignment RA for $S(G)$, there exists a canonical range assignment RA_c such that $c(RA_c) \leq c(RA)$.

Proof. We prove that any symmetric connecting noncanonical range assignment RA can be transformed into a feasible canonical range assignment RA_c through a sequence of iterative steps, where each step does not increase the cost of the range assignment. Every step considers a node u whose transmitting range is not canonical, and derives a symmetric connecting range assignment such that the transmitting range for u is canonical. Since the number of noncanonical points in $S(G)$ decreases at each step, the iterative process ends in finite time.

We describe the generic step of this process. Let v be a noncanonical point and let $RA(v)$ be its transmitting range. We have the following cases:

1. $RA(v) < \gamma\lambda$. In this case, the transmitting range of v is not sufficiently large to reach points in the YZ-components of other gadgets. Note that if $RA(v) < \lambda + \varepsilon$ then v cannot be a bridge between the YZ-component and the VX-component. Hence, its transmitting range can be decreased to λ or λ' (depending on whether $v \in V \cup X$ or $v \in Y \cup Z$) without disconnecting the graph and preserving symmetry. Assume now $RA(v) \geq \lambda + \varepsilon$. Assume without loss of generality that $v \in g_{ab}$, for some $(a, b) \in E$. If $v \in V_{ab} \cup Y_{ab}$, then its transmitting range can be reduced to $\lambda + \varepsilon$ without disconnecting the graph and preserving symmetry. Otherwise, consider the range assignment RA_{ab} such that

   - $RA_{ab}(w) = RA(w)$ for any $w \in S(G) - g_{ab}$;
   - $RA_{ab}(a) = RA_{ab}(y_{ab}) = \lambda + \varepsilon$;
   - $RA_{ab}(b) = \lambda$ and $RA_{ab}(y_{ab}) = \lambda'$;
   - $RA_{ab}(x) = \lambda$ for any $x \in X_{ab}$;
   - $RA_{ab}(z) = \lambda'$ for any $z \in Z_{ab}$.

   Given the properties of points in a gadget, it follows that RA_{ab} is symmetric. Furthermore, the communication graph resulting from RA_{ab} is connected, and RA_{ab} is canonical in g_{ab}, and hence, in v. Let $c(S(G)\backslash g_{ab}) = \sum_{v \in S(G)\backslash g_{ab}} RA(v)^2$. Given the requirement for symmetry, we have

   $$c(RA) \geq c(S(G)\backslash g_{ab}) + 2 \cdot RA(v)^2 + (l_1 + 1) \cdot \lambda^2 + (l_2 + 1) \cdot \lambda'^2,$$

   where $l_1 = |X_{ab}|$ and $l_2 = |Z_{ab}|$.

   On the other hand, we have

   $$c(RA_{ab}) = c(S(G)\backslash g_{ab}) + 2 \cdot (\lambda + \varepsilon)^2 + (l_1 + 1) \cdot \lambda^2 + (l_2 + 1) \cdot \lambda'^2.$$

   Since $RA(v) \geq \lambda + \varepsilon$, we can conclude that $c(RA) - c(RA_{ab}) \geq RA(v)^2 - (\lambda + \varepsilon)^2 \geq 0$.

2. $RA(v) \geq \gamma\lambda$. In this case v could be a bridge point between many YZ-components and VX. Assume without loss of generality that $v \in g_{ab}$, for some $(a, b) \in E$. We first transform the range assignment as described above, obtaining the range assignment RA_{ab}, with $c(RA) - c(RA_{ab}) \geq \gamma^2\lambda^2 - (\lambda + \varepsilon)^2$. However, RA_{ab} in general is not symmetric and could leave some YZ-components isolated. For this reason, we consider the isolated components $YZ_1, \ldots, YZ_k$ in the graph generated by RA_{ab}, and for each of this component we apply the same construction as for the gadget g_{ab}. The resulting range assignment $R\bar{A}_{ab}$ is symmetric and connecting. In order to prove the lemma, we have to show that $c(R\bar{A}_{ab}) \leq c(RA)$. Note that the cost associated with any YZ-component YZ_i could increase in the new range assignment $R\bar{A}_{ab}$. However, observing that because of symmetry at least one node in YZ_i must have transmitting range of at

least $\gamma\lambda$, the increase for every YZ-component is bounded by $2(\lambda + \varepsilon)^2 - (\gamma\lambda)^2 - \lambda^2$. Considering that $k \le m - 1$, we have

$$c(RA) - c(R\bar{A}_{ab}) \ge \gamma^2\lambda^2 - (\lambda + \varepsilon)^2 - (m - 1)(2(\lambda + \varepsilon)^2 - (\gamma\lambda)^2 - \lambda^2) \ge 0$$

by inequality (7.1). This ends the proof of the lemma.

Consider a planar cubic graph G and a planar orthogonal drawing $D(G)$ of G, and let $2h$ be the number of nodes added in the second step of the construction. Assign to every node in G and in $D(G)$ a weight equal to its degree. The following lemma relates the cost of a vertex cover for G with that of a vertex cover for $D(G)$.

Lemma 7.4.5 *Let G be a planar cubic graph, $D(G)$ a planar orthogonal drawing of G, and let $2h$ be the number of nodes added in the construction. Assign to every node in G and in $D(G)$ a weight equal to its degree. Then, G has a vertex cover of cost $\le k$ if and only if $D(G)$ has a vertex cover of cost $\le k + 2h$.*

Proof. The proof follows easily from Lemma 3.1 of (Kirousis et al. 2000) and by observing that every node added in the construction of $D(G)$ has degree 2.

We are now ready to prove that SRA in two-dimensional networks is NP-hard.

Theorem 7.4.6 *Solving the SRA problem in two- and three-dimensional networks is NP-hard.*

Proof. We show a polynomial time reduction from MINWEIGHTEDVERTEXCOVER for planar cubic graphs.

Let G be any planar cubic graph and let $D(G) = (V, E)$ be a straight-line planar orthogonal drawing of G. By Lemma 7.4.5, the problems of determining a vertex cover of minimum weight on $D(G)$ and on G are equivalent, and hence MINWEIGHTEDVERTEXCOVER on $D(G)$ is NP-hard.

Consider now the set of two-dimensional points $S(G)$ obtained by constructing gadgets on every edge of $D(G)$ as described above. In Clementi et al. (1999), it is proved that for any $D(G)$ it is possible to derive $S(G)$ in polynomial time, and that points in the gadgets satisfy condition $(a)\ldots(e)$, for positive constants γ, λ, λ', ε such that $\lambda + \varepsilon > \lambda'$ and

$$(\gamma\lambda)^2 > (m - 1)\left((\lambda + \varepsilon)^2 - \lambda^2\right) + (\lambda + \varepsilon)^2$$

$$> \frac{m - 1}{m}\left((\lambda + \varepsilon)^2 - \lambda^2\right) + (\lambda + \varepsilon)^2.$$

It can be seen that the same choice of γ, λ, λ', and ε enables points in the gadgets to have the additional properties $(b')\ldots(d')$.

Let $Y = \bigcup_{a,b\in E} Y_{ab}$, $X = \bigcup_{a,b\in E} X_{ab}$, $Z = \bigcup_{a,b\in E} Z_{ab}$, and $V = \bigcup_{a,b\in E} V_{ab}$. We now prove that $D(G)$ has a vertex cover C of cost k if and only if $S(G)$ has a feasible range assignment of cost $(|X| + |V| - |C|)\lambda^2 + (|Y| + |Z| - k))\lambda'^2 + (|C| + k)(\lambda + \varepsilon)^2$.

Assume $D(G)$ has a vertex cover C of cost k. Consider the canonical range assignment RA_c on $S(G)$ where

- $RA_c(v) = \lambda + \varepsilon$ for any $v \in C$,

- $RA_c(v) = \lambda$ for any $v \in V \backslash C$,

and transmitting ranges of points in Y are assigned in such a way that symmetry is guaranteed. Since C is a cover of $D(G)$, RA_c is connecting, and has cost $(|X| + |V| - |C|) \lambda^2 + (|Y| + |Z| - k))\lambda'^2 + (|C| + k)(\lambda + \varepsilon)^2$.

Consider now any symmetric connecting range assignment RA for $S(G)$. By Lemma 7.4.4, there exists a canonical range assignment RA_c such that $c(RA_c) \leq c(RA)$. Hence, we can restrict our attention to the canonical range assignment RA_c. Let k be the number of points in Y whose transmitting range is $\lambda + \varepsilon$, and let C be the set of points in V whose transmitting range is $\lambda + \varepsilon$. The cost of RA_c is $(|X| + |V| - |C|) \lambda^2 + (|Y| + |Z| - k))\lambda'^2 + (|C| + k)(\lambda + \varepsilon)^2$. Since RA_c is canonical, it follows that C is a vertex cover of $D(G)$ and that, because of symmetry, the cost of C is k. This ends the proof of the theorem.

7.4.3   Approximation algorithms for WSRA

Since solving WSRA in two- and three-dimensional networks is NP-hard (this follows from the fact that a restricted version of WSRA, SRA, is NP-hard), some authors have studied approximation algorithms for WSRA.

First, we observe that the range assignment RA_T used in Theorem 7.3.2 to approximate RA within a factor of 2 generates a communication graph whose symmetric subgraph is connected. So, RA_T is a 2-approximate solution of WSRA also. Better polynomial time approximation algorithms for WSRA have been introduced in (Althaus et al. 2003): the first algorithm has an approximation ratio of $\frac{5}{3} + \varepsilon$, for any positive constant $\varepsilon > 0$, while the second, which is more computationally efficient, has an approximation ratio of $\frac{11}{6}$. Further, the authors of (Althaus et al. 2003) present an exact branch and cut algorithm for solving WSRA based on an integer linear program formulation of the problem. Experimental results show that the branch and cut algorithm solves instances with up to 35–40 nodes (with randomly generated positions) in 1 hour. Most importantly, the experimental results show that the average improvement of the exact solution over RA_T, which can be easily calculated, is in the range 4–6%. This means that the average case approximation factor of RA_T is much smaller than its worst-case factor of 2.

7.5   The Energy Cost of the Optimal Range Assignment

In the previous section, we have considered two restricted versions of the RA problem, and we have seen that, in case of two- and three-dimensional networks, the computational complexity of the restricted problems does not change with respect to the general case. It is also interesting to evaluate which is the impact of the stronger connectivity requirements

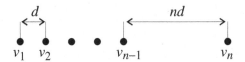

Figure 7.6 Problem instance for which $\frac{c_S}{c_{RA}} \in \Omega(n)$.

on the cost of the optimal range assignment. In other words, denoting with c_{RA}, c_{WS} and c_S the energy cost of the optimal solutions of RA, WSRA, and SRA for a certain problem instance, which is the relation between these costs?

Clearly, we have $c_{RA} \leq c_{WS} \leq c_S$, since WSRA and SRA are increasingly constrained versions of RA. By observing that the range assignment RA_T used in Theorem 7.3.2 to approximate RA within a factor of 2 generates a communication graph whose symmetric subgraph is connected, it follows that $\frac{c_{WS}}{c_{RA}} \in O(1)$ (Blough et al. 2002). As for c_S and c_{RA}, the following construction shows that $\frac{c_S}{c_{RA}} \in \Omega(n)$ already in one-dimensional networks.

The construction is as follows (Blough et al. 2002). We have n colinear nodes $v_1, \ldots, v_n$ (nodes are ordered from left to right). Nodes v_i, v_{i+1} are at distance $d > 0$, for $i = 1, \ldots, n - 2$; the rightmost node v_n is at distance nd from node v_{n-1} (see Figure 7.6). By connecting each node to its closest left and right neighbor, we obtain a range assignment $\overline{RA}$ that is connected and has cost equal to $c(\overline{RA}) = (n - 2)d^\alpha + 2(nd)^\alpha$. So, the optimal solution to RA has at most cost equal to $c(\overline{RA})$. If the range assignment must satisfy (strong) symmetry, the fact that $RA(v_{n-1}) \geq nd$ (this is necessary to connect node v_{n-1} and node v_n) implies that node v_{n-1} has a unidirectional link to every node v_i, with $i = 1, \ldots, n - 2$. Given strong symmetry, this implies that $RA(v_i) \geq nd$ for any node v_i, that is, $c_S \geq n(nd)^\alpha$. Then, we have $\frac{c_S}{c_{RA}} \in \Omega(n)$.

In other words, the results above prove that the requirement for weak symmetry has only a marginal effect on the energy cost of the range assignment, while it eases significantly the integration of topology control mechanisms with existing protocols (e.g. routing and MAC). On the other hand, imposing the stronger requirement of symmetry results in a considerable additional energy cost. Overall, we can conclude that *weak symmetry is a desirable property of the range assignment*.

8

Energy-efficient Communication Topologies

In the previous chapter, we have considered the problem of computing a set of transmitting range assignments of minimum energy cost that generates a strongly connected communication graph. Considerable research has been devoted also to the problem of identifying good topologies for energy-efficient communication among the network nodes. In other words, we are given the communication graph G obtained when all the nodes transmit at maximum power, and the goal is to identify a sparse subgraph G' of G such that only energy-efficient links are retained in G'. The criterion used to determine whether a certain link in G is energy efficient depends on the communication pattern considered. Typically, the focus is on either end-to-end communication between arbitrary nodes (unicast) or on one-to-all communications (broadcast). In this chapter, we first analyze this topology optimization problem for unicast communications, and then we consider the same problem in case of broadcast. Note that all the results presented in this chapter are for two-dimensional networks.

8.1 Energy-efficient Unicast

Let $G = (N, E)$ denote maxpower graph, that is, the communication graph that is generated when all the nodes transmit at maximum power. In what follows, we assume that G is connected.

Let P_{uv} be any directed path connecting nodes u and v in G. The *power cost* of $P_{uv} = \{u = w_0, w_1, \ldots, w_h, w_{h+1} = v\}$ is defined as the sum of the power costs of the single edges, that is,

$$pc(P_{uv}) = \sum_{i=0}^{h} \delta(w_i, w_{i+1})^\alpha,$$

where α is the distance-power gradient. The path in G connecting u and v and consuming the minimum power is denoted $P_{uv}^{\min, G}$, and is called the *minimum-power path* between u

and v in G. If the minimum-power path between u and v is not unique, we refer to any of these paths as the minimum-power path.

Definition 8.1.1 (Power stretch factor) *Let G' be an arbitrary subgraph of the maxpower graph $G = (N, E)$. The power stretch factor of G' with respect to G, denoted as $\rho_{G'}$, is the maximum over all possible node pairs of the ratio between the power cost of the minimum-power path in G' and in G. Formally,*

$$\rho_{G'} = \max_{u,v \in N} \frac{pc(P_{uv}^{\min,G'})}{pc(P_{uv}^{\min,G})}.$$

By convention, we define $\rho_{G'} = \infty$ if there exist nodes u, v that are connected in G, but are disconnected in G'.

The power stretch factor is a generalization of the concept of *distance stretch factor*, which is well known in computational geometry (Goodman and O'Rourke 1997). Another similar concept is the *hop stretch factor*, which measures the ratio of the hop counts rather than that of power or distance. An example of maxpower graph G, of a subgraph G' of G, and the correspondent power, distance, and hop stretch factors are reported in Figure 8.1.

Definition 8.1.2 (Power spanner) *Let $G = (N, E)$ be the maxpower graph, with $|N| = n$. A subgraph G' of G is said to be a power spanner of G if $\rho_{G'} \in O(1)$.*

Similar definitions can be given for distance and hop spanners.

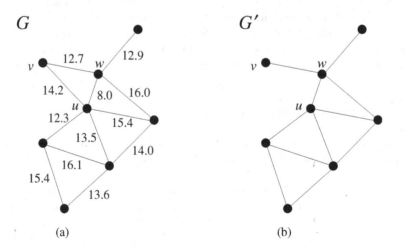

Figure 8.1   The graph (a) is the maxpower graph G, where the edges are labeled with their length. The subgraph G' of the maxpower graph obtained by removing edge (u, v) is reported in (b). The power stretch factor of G' is 1 (we assume $\alpha = 4$), since edge (u, v) is energy inefficient, and the alternative path $\{u, w, v\}$ is used also in the maxpower graph. The distance stretch factor of G' is $\frac{20.7}{14.2} = 1.46$. The hop stretch factor is 2, because the minimum-hop path connecting nodes u and v in G' is the two-hops path $\{u, w, v\}$.

In general, we would like to identify a subgraph G' (also called *routing graph* in the following) of the maxpower graph G that has a low power stretch factor (possibly, a power spanner of G), and which is significantly sparser than the original graph. This routing graph can be seen as the input to the routing protocol, which computes the routes between nodes considering only the links in G'. Given the power spanning property, we have the guarantee that the power needed to communicate along these routes is 'almost minimal'. The advantage of using G' instead of G is that the routing overhead (which is mainly due to the limited flooding of control messages in the route discovery phase[1]) is reduced if G' is considerably sparser than G.

Note that in this approach to the topology control problem it is implicitly assumed that a node changes the transmit power on a per-packet basis: when node u has to send a packet to node v, it sets the transmit power level to the minimum value needed to reach the next node in the route to v. Thus, according to the taxonomy introduced in Chapter 3, we can classify the results presented in this section as per-packet topology control.

Besides being a sparse[2] power spanner, other desirable features of the routing graph have been identified. In particular, the node degree in the constructed topology should be upper bounded by a constant. Note that the fact that G' is sparse guarantees that the *average*, and not the *maximum*, node degree in the graph is constant. Having an upper bound on the maximum node degree is desirable to avoid bottlenecks in the communication graph. If the routing graph is used in conjunction with a geographic routing protocol (such as the protocols presented in (Bose et al. 2001; Karp and Kung 2000; Khun et al. 2003)), then planarity is a fundamental property to guarantee message delivery. Finally, and most importantly, the routing graph should be constructed in a localized and fully distributed fashion. In other words, any node u in the network should be able to compute its local view of G' (i.e. the edges of G' incident in it) on the basis of only the information regarding nodes that are immediate (or, at most, two-hops) neighbors of u in the maxpower graph.

Summarizing, the routing graph G' should

- be a power spanner of the maxpower graph;

- be sparse;

- have bounded node degree;

- be planar;

- be easily computable in a fully distributed and localized fashion.

Several routing graphs that satisfy some or all of the requirements above have been proposed in the literature. Most of them are based on subgraphs of G that are known to be good distance spanners. In fact, it can be easily seen that if a certain routing graph G' is a distance spanner of graph G, then it is also a power spanner of G (note that the reverse implication in general is not true). Thus, the considerable body of research devoted to distance spanners in computational geometry can be used to design good routing graphs (Goodman and O'Rourke 1997).

---

[1] This is true for reactive routing protocols, which are known to perform very well in ad hoc/sensor networks.

[2] We recall that, in graph theoretic terms, a graph on n nodes is sparse if the number of edges in it is $O(n)$ (see Appendix A).

Table 8.1   Distance stretch factor, power stretch factor, average and maximum node degree of different proximity graphs

	Distance	Power	Avg. Degree	Max. Degree
RNG	$n-1$	$n-1$	$O(1)$	$n-1$
GG	$\frac{4\pi\sqrt{2n-4}}{3}$	1	$O(1)$	$n-1$
RDT	$\frac{1+\sqrt{5}}{2}\pi$	$\left(\frac{1+\sqrt{5}}{2}\pi\right)^{\alpha}$	$O(1)$	$\Theta(n)$
YG_c	$\frac{1}{1-2\sin\frac{\pi}{c}}$	$\frac{1}{1-(2\sin\frac{\pi}{c})^{\alpha}}$	$O(1)$	$n-1$

In particular, the following graphs (for a definition of these graphs see Appendix A) borrowed from computational geometry have been used to build good routing graphs for ad hoc networks: the Relative Neighborhood Graph (RNG), the Gabriel Graph (GG), the Delaunay Triangulation (DT), and the Yao Graph of parameter c (YG_c). These graphs are called *proximity graphs*, since the set of links incident in any node u of the computed graph can be calculated on the basis of the position of the neighbor nodes in the maxpower graph. Thus, proximity graphs can be constructed in a fully distributed and localized way.

The following relationships between proximity graphs have been proven (Goodman and O'Rourke 1997; Li et al. 2002): for any set of points N, $RNG(N) \subseteq GG(N)$, and $RNG(N) \subseteq YG_c(N)$, for any $c \geq 6$. Furthermore, $MST(N)$ is contained in $RNG(N)$, $GG(N)$, $DT(N)$, and $YG_c(N)$, for any $c \geq 6$. The distance and power spanning ratios of these graphs are summarized in Table 8.1, along with the average and maximum node degree. In the table, RDT is the *restricted* Delaunay Triangulation, where the edges exceeding the nodes' maximum transmitting range are removed (see (Gao et al. 2001)).

As seen from Table 8.1, the GG has optimal power stretch factor. The algorithm reported in Figure 8.2, which was presented in (Song et al. 2004), can be used to compute the GG in a fully distributed and localized way. The algorithm relies on the assumption that every node in the network knows its position on the plane. This can be accomplished by equipping nodes with low-power GPS receivers, or by using other location estimation techniques.

We recall that an edge $(u, v) \in G$ is included in the Gabriel Graph if and only if the disk with the same edge as diameter contains no node of G (see Figure 8.3). It is immediate to see that the algorithm reported in Figure 8.2 constructs the GG. Considering that the maximum node degree in the maxpower graph can be as high as $n - 1$, it can be easily seen that the algorithm has $O(n^2)$ time complexity. The message complexity is $O(n)$, since every node in the graph transmits a single message.

Although the GG has optimal power stretch factor, its maximum node degree can be as high as $n - 1$. The same holds for the other graphs listed in Table 8.1. For this reason, several variants of the above proximity graphs have been proposed, with the purpose of having a constant bound on the maximum node degree. Unfortunately, it has been shown that no geometric graph with constant node degree contains the minimum-power path for any pair of nodes (Wang et al. 2002). Thus, no power-optimal spanner with a constant bounded maximum node degree exists. To date, the routing graph with constant maximum node degree that has the best power stretch factor is the *OrdYaoGG* graph of (Song et al. 2004), which is obtained by building the YG_c graph, with $c > 6$, on the top of the GG. The *OrdYaoGG* graph has power stretch factor of $\rho = \frac{1}{1-(2\sin\frac{\pi}{c})^{\alpha}}$, and maximum node degree

Algorithm GABRIELGRAPH:
(*Algorithm for node u*)

ID_u is the identifier of node u, and (x_u, y_u) is its location
$E_G(u)$ and $E_{GG}(u)$ are the set of links of the maxpower graph and of the *GG* node u
                is aware of
$disk(u, v)$ is the disk with edge (u, v) as diameter

1. Initialization
      Locally broadcast message $(ID_u,(x_u, y_u))$ at maximum power
      $E_G(u) = E_{GG}(u) = \emptyset$

2. Processing of incoming messages
      upon receiving message $(ID_v,(x_v, y_v))$ from node v
         add (u, v) to $E_G(u)$
         check whether exists edge $(u, w) \in E_G(u)$ such that $w \in disk(u, v)$
         if *no*, then add (u, v) to $E_{GG}(u)$
         for each $(u, w) \in E_{GG}(u)$
            check whether $v \in disk(u, w)$
            if *yes*, then remove (u, w) from $E_{GG}(u)$

3. Termination
      after processing all incoming messages, $E_{GG}(u)$ contains all the edges of *GG*
      incident in u

Figure 8.2 Algorithm for constructing the Gabriel Graph.

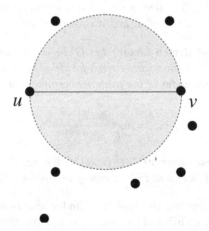

Figure 8.3 Edge (u, v) is included in the *GG* if and only if the disk with the same edge as diameter (shaded area) contains no node.

of $c + 5$, where $c > 6$ is the parameter of the Yao graph. For example, setting $c = 9$ and $\alpha = 2$, we have a power stretch factor of 1.88 with a bound on the maximum node degree of 14. Although *OrdYaoGG* can be constructed in a fully distributed and localized fashion, its computation requires the exchange of a considerable number of messages ($24\,n$ messages in the worst case). For this reason, the authors of (Song et al. 2004) proposed the *SYaoGG* graph, which is a simplified version of *OrdYaoGG* that can be constructed exchanging at most $3\,n$ messages. The power stretch factor of this graph is $\rho = \frac{\sqrt{2}^{\alpha}}{1-(2\sqrt{2}\sin\frac{\pi}{c})^{\alpha}}$, and the maximum node degree is c, where $c > 8$ is the parameter of the Yao graph. Setting $c = 9$ and $\alpha = 2$ as above, we have a power stretch factor of 31.16 and a maximum node degree equal to 9.

8.2 Energy-efficient Broadcast

Another problem that has been considered in the literature is the determination of topologies for energy-efficient broadcast: we are given the maxpower graph G, and the goal is to determine a sparse subgraph G' of G (the *broadcast graph*) such that broadcasting in G' is almost as energy efficient as broadcasting in the maxpower graph. Here, the advantage of using a graph that is sparser than the maxpower graph as the broadcasting topology is that the well-known broadcast storm phenomenon (Ni et al. 1999) can be reduced. This phenomenon occurs when many nodes in a neighborhood try to relay the broadcast message at the same time, resulting in serious redundancy, bandwidth contention, and collision.

Before presenting the results, we introduce the concept of *broadcast stretch factor*. Let us consider the maxpower communication graph G. Any broadcast generated by node u can be seen as a directed spanning tree T of G rooted at u, which we call a *broadcast tree*. The power cost of the broadcast tree T is defined as follows. Denoting with $pc_T(v)$ the power consumed by node v to broadcast the message along T, we have that $pc_T(v) = 0$ if v is a leaf node of T, and $pc_T(v) = \max_{(v,w) \in T} \delta(v, w)^{\alpha}$ otherwise. The total power needed to broadcast the message along the broadcast tree T is then $pc(T) = \sum_{v \in N} pc_T(v)$. We call this cost the *power cost* of T. The tree in G rooted at u with minimum power cost is called the *minimum-power broadcast tree* of u, and it is denoted $T_u^{\min, G}$.

Definition 8.2.1 (Broadcast stretch factor) *Let G' be an arbitrary subgraph of the maxpower graph $G = (N, E)$. The broadcast stretch factor of G' with respect to G, denoted as $\beta_{G'}$, is the maximum over all nodes of the ratio between the minimum-power broadcast tree in G' and in G. Formally,*

$$\beta_{G'} = \max_{u \in N} \frac{pc(T_u^{\min, G'})}{pc(T_u^{\min, G})}.$$

Definition 8.2.2 (Broadcast spanner) *Let $G = (N, E)$ be the maxpower graph, with $|N| = n$. A subgraph G' of G is said to be a broadcast spanner of G if $\beta_{G'} \in O(1)$.*

Similar to the case of unicast, the goal is to find sparse broadcast spanners of G that can be computed in a fully distributed and localized fashion. Unfortunately, this task turns out to be more difficult than in the case of unicast communication.

The main difficulty arises from the fact that the computing of the minimum-power broadcast tree rooted at a certain node is a NP-hard problem (Cagali et al. 2002; Liang 2002), under the hypothesis that the nodes can use a set of discrete power levels $\{P_1, \ldots, P_k\}$. So, directly computing the broadcast stretch factor of a certain subgraph G' of G is virtually impossible, since this would require solving a NP-hard problem.

Given this hardness result, several authors have presented heuristic solutions that approximate the optimal (but hard to compute) solution of the minimum-power broadcast tree problem. One popular such heuristic, which we present in the following, is the Broadcast Incremental Power (BIP) algorithm introduced in (Wieselthier et al. 2000).

The BIP algorithm, which is reported in Figure 8.4, is a variation of the well-known Prim's algorithm for finding the MST. The algorithm starts by finding the node that the source node u can reach with minimum power cost. This node is added in the set C of covered nodes, that is, the nodes that have received the broadcast message. At the generic step i, BIP considers all the uncovered nodes and, for any such node v, calculates the incremental cost of adding v to the current spanning tree. The node $\overline{v}$ with minimum incremental cost is added to the set of covered nodes, and is now part of the spanning tree. This process is repeated until all the nodes in the network are covered.

In (Wan et al. 2002), Wan et al. showed that the approximation ratio of BIP is between $\frac{13}{3}$ and 12. Then, a broadcast spanner of G can be constructed as follows: for any node u in G, apply the BIP algorithm to construct the broadcast tree T_u rooted at u. Let $G_{BIP} = \bigcup_{u \in N} T_u$, that is, link (u, v) is in G_{BIP} if and only if it is contained in one of the broadcast trees computed by BIP. Given that BIP approximates the optimal solution (which is built on the maxpower graph) by a factor at most 12, it follows that, for any $u \in N$, G_{BIP} contains a

---

Algorithm BROADCASTINCREMENTALPOWER:

u is the source node
C is the set of currently covered nodes
T is the current spanning tree
N is the set of network nodes

1. Initialization
   $C = \{u\}$
   $T = \{u\}$

2. Repeat until $C = N$
   for each node $v \in N - C$, compute the incremental cost $ic(v)$ of adding v to T
   let $\overline{v}$ be the node in $N - C$ with minimal incremental cost
   $C = C \bigcup \{\overline{v}\}$
   add $\overline{v}$ in the current spanning tree T

3. Termination
   when $C = N$, T is a broadcast spanning tree rooted at u

Figure 8.4 BIP algorithm for constructing the broadcast tree.

broadcast tree rooted at u of cost at most $O(1)$ times the cost of the optimal tree; that is, G_{BIP} is a broadcast spanner of G. Unfortunately, graph G_{BIP} in general is not sparse. Furthermore, BIP is a *centralized* algorithm that requires *global knowledge* (it requires knowing at least the set N of network nodes).

Another graph that can be used to construct broadcast spanners is the MST, which approximates the minimum-power broadcast tree by a factor between 6 and 12 (Wan et al. 2002). Unfortunately, the computation of the MST also requires global knowledge. In order to circumvent this problem, Li et al. (Li et al. 2004) have recently proposed a localized, fully distributed algorithm that constructs a local approximation of the MST. The algorithm, called $LMST_k$, requires exchanging $O(n)$ messages (although the hidden constant is larger than 225), and builds a $O(n^{\alpha-1})$ approximation of the energy-optimal broadcast tree. Thus, $LMST_k$ cannot be used to compute a broadcast spanner of G.

Summarizing, the problem of designing a distributed and localized algorithm that can be used to build a broadcast spanner of G is still open.

Before ending this section, we want to outline the similarities between the range assignment problem discussed in Chapter 7 and the problem of energy-efficient broadcast. Suppose G is the maxpower graph on the set of points N. In the. RA problem, the goal is to find the energy-optimal range assignment that generates a connected communication graph. Suppose an arbitrary node $u \in N$ wants to broadcast a message m, and let RA be the optimal range assignment. A very simple broadcast scheme is flooding: node u transmits m at distance $RA(u)$, and every other node v, upon receiving m for the first time, retransmits it at distance $RA(v)$. It is immediately seen that after all nodes in N have transmitted the message once m has been broadcast throughout the network. Thus, *the energy cost of RA is an upper bound to the power cost of any broadcast tree in G*. We recall that the energy cost of the optimal range assignment (and of the optimal weakly symmetric range assignment) differs from the cost of the MST at most for a factor 2. Since the MST is a broadcast spanner of G, this implies that the communication graph generated by the optimal (weakly symmetric) range assignment is a broadcast spanner of G. Unfortunately, this does not help very much, since computing this graph in two and three-dimensional networks is NP-hard.

Part IV

Distributed Topology Control

9

Distributed Topology Control: Design Guidelines

In Part III of this book, we have considered several topology optimization problems in which we are given the node locations, and the goal is to identify 'optimal' network topologies with respect to a certain metric (e.g. energy cost, energy-efficient unicast/broadcast, and so on). In this approach to topology control, the emphasis is on the properties of the generated topology, rather than on the process needed to build such topology. Typically, it is assumed that all the information regarding node positions is available to a central entity, which uses this information to compute the 'optimal' topology. In this and in the next chapters, we consider more practical approaches to the topology control problem, in which the challenge is to design lightweight, fully distributed protocols that generate a 'reasonably good' topology. Here, the focus is then on the topology generation process, rather that on the 'quality' of the resulting topology.

We start by discussing the guidelines that should be used in the design of topology control protocols. We then consider three approaches to distributed topology control, which are ordered by the amount of information available to the network nodes: first, we consider location-based protocols, in which every node knows its own location (Chapter 10); then, we proceed to direction-based protocols, in which nodes do not know their position, but they can estimate the direction of neighbor nodes (Chapter 11); finally, we present neighbor-based topology control, in which nodes only know the number and identity of their neighbors (Chapter 12). The last chapter of Part IV of this book deals with node mobility, discussing how mobility affects the implementation and the performance of topology control protocols.

9.1 Ideal Features of a Topology Control Protocol

Which are the features that a topology control protocol should ideally have?

Given that the lack of a centralized authority (such as the base station in a cellular network) is one of the peculiar features of ad hoc and sensor networks, centralized approaches are doomed to perform poorly in realistic application scenarios. For this reason, only solutions that can be implemented in a *fully distributed* and *asynchronous* fashion have some practical relevance.

Another important feature of a protocol for topology control is *locality*, which refers to the nodes' ability to build their view of the network topology by using local information only, that is, information regarding up to h-neighbors in the maxpower graph, where h is a small constant (2–3 at most). Localized solutions have several advantages with respect to approaches that require networkwide information exchange: since network nodes can build their local view of the topology by exchanging few messages with neighbor nodes only, localized protocols can be classified as lightweight solutions, which can be implemented in very large networks also; furthermore, locality implies that the network topology can be easily reconfigured when nodes leave/join the network, or in presence of node mobility.

As mentioned above, the goal of a distributed topology control protocol is to build a 'reasonably good' topology. But what do we mean by 'reasonably good'? What are the essential features that the generated topology should have? For the reasons discussed in Section 7.4, *the topology generated* by the topology control protocol *should rely on bidirectional links only*. Furthermore, it is desirable that the topology control protocol *preserves connectivity*. In other words, if the network is connected when all the nodes communicate with maximum transmit power (i.e. if the maxpower graph is connected), then it should preserve this property also after every node in the network has executed the topology control protocol. So, only redundant links should be removed from the network topology.

In the following, we will see that there is a considerable difference between requiring connectivity preservation in the *worst case* (i.e. for any node placement, if the maxpower communication graph is connected, then the network remains connected after the execution of the topology control protocol) and requiring connectivity *with high probability*. As we shall see, the former property can be achieved only if the physical node degree in the network is unbounded (the notion of physical node degree is introduced in Section 9.3), while the latter can also be achieved when the number of physical neighbors of a node are upper bounded. As discussed in Section 9.3, building a network topology in which nodes have small physical degree is highly desirable, since this parameter is a measure of the interference generated by a transmitting node: if the physical degree of node u is small, the number of nodes impacted by u's transmission is relatively small, and spatial reuse is increased. So, we can state that at least two of the desired topological properties discussed above are conflicting: ensuring worst-case connectivity and generating a topology with small physical node degree.

A final aspect to consider in the design of a topology control mechanism is the quality of information required by the protocol: since obtaining very accurate information such as node locations is, in general, quite expensive (in terms of additional hardware required on the nodes, or message overhead, or both), it is desirable that the protocol relies on 'low-quality' information. This issue is carefully discussed in Section 9.2.

Summarizing, a topology control protocol should

1. be fully distributed and asynchronous;

2. be localized;

3. generate a topology that preserves the original network connectivity and relies on bidirectional links only;

4. generate a topology with small physical node degree;

5. rely on 'low-quality' information.

9.2 The Quality of Information

An important aspect to be considered in the design of topology control protocols is the type of information used by the nodes to build the local view of the topology: nodes can use high-quality information (e.g. neighbor node locations), medium-quality information (e.g. directional information, or distance to neighbors), or low-quality information (number and identity of neighbor nodes). In general, there is a direct relationship between information quality and energy efficiency of the computed topology: the more accurate the information available to the nodes, the more energy savings can be achieved. However, information quality (and, thus, energy savings) must be carefully traded off with the *cost* incurred for making the information available to the nodes. The cost is due to either some additional hardware required on the nodes (e.g. low-power GPS receivers in case of location information) or the message overhead needed to produce/update high-quality information, or both.

We clarify this point with an example. Suppose protocol P_1 is based on location information and protocol P_2 is based on distance estimation. Clearly, the cost of implementing P_2 in a real network is significantly lower than that required by P_1, since estimating distance between nodes requires cheaper hardware and/or less message exchange than does estimating node positions. So, if the energy savings provided by protocol P_1 are not considerably higher than those achieved by P_2, a solution based on protocol P_2 may be preferable in practice.

Another argument in favor of designing protocols based on low-quality information is that they can be used in a wider range of application scenarios. For instance, it is well known that location estimation techniques perform poorly in indoor environments because of the hardly predictable propagation of the radio signal. So, location-based topology control protocols such as those presented in Chapter 10 cannot be used in this case.

9.3 Logical and Physical Node Degrees

As discussed in Chapter 3, one of the motivations for topology control is its potential to reduce interference between concurrent transmissions. A typical measure used to quantify the expected interference is the node degree of the communication graph: if the transmitting node u has small degree, relatively few nodes will experience interference during u's transmission. For this reason, it is desirable to generate topologies with small average node degree. However, a clarification about the term 'node degree' is in order.

Most of the current literature on topology control defines node degree as follows:

Definition 9.3.1 (Logical node degree) *Let $G_P = (N, E_P)$ be the communication graph generated by a certain topology control protocol P. For a given $u \in N$, the (logical) node degree of u in G_P is defined as*

$$L\,Deg(u) = |\{v \in N : (u, v) \in E_P\}|.$$

The definition above is the traditional definition of node degree used in graph theory (see Appendix A). Since in the distributed setting typical of ad hoc and sensor networks every node independently builds its local view of the communication graph, we can equivalently define the logical degree of a certain node u as the number of nodes in its neighbor list after the protocol execution.

Note that what is defined above is a *logical node degree*, since some of the nodes that are within u's transmitting range might not appear in u's neighbor list (typically, because their link to node u is not energy efficient). As a consequence, the logical node degree is not very effective in measuring the expected interference experienced by nodes: if a certain node v is within u's transmitting range but is not listed in u's neighbor list (i.e. edge (u, v) is not in E_P), it will not contribute to u's logical degree, but it is definitely affected by node u's transmission.

To circumvent this problem, the authors of (Blough et al. 2003b) have introduced the notion of *physical node degree*, which is defined as follows:

Definition 9.3.2 (Physical node degree) *Let $G_P = (N, E_P)$ be the communication graph generated by a certain topology control protocol P. For a given $u \in N$, the (physical) node degree of u in G_P is the number of nodes within u's range when it transmits with the minimum power needed to reach its farthest neighbor in G_P, that is, when u transmits at the broadcast power. Formally,*

$$P\,Deg(u) = |\{v \in N : \delta(u, v) \leq \max_{(u,w)\in E_P} \delta(u, w)\}|.$$

It is easy to see that the logical degree is a lower bound to the physical degree.

To clarify the difference between logical and physical node degrees, let us consider the example shown in Figure 9.1. The figure reports a sample topology built by the CBTC protocol introduced in (Wattenhofer et al. 2001), which will be presented in detail in Chapter 11. After CBTC's execution, node u has nodes v, w, x, y, z in its neighbor list, that is, it has logical degree 5. However, when u transmits a message to its furthest neighbor, node w, more than 5 nodes are within its radio range. Consider for instance node t; this node does not have a direct link to u in the constructed topology, because communicating along the two-hops' path $\{t, v, u\}$ is more energy efficient than the direct communication from t to u. Nevertheless, t is affected by u's transmission to node w, and it contributes to its physical degree. Note that the physical degree can be substantially higher than the logical degree: in the example of Figure 9.1, the logical degree of node u is 5, while its physical degree is 10.

The following theorem proves a fundamental trade-off between worst-case connectivity and physical node degree:

Theorem 9.3.3 *Let $G = (N, E)$ be the maxpower communication graph on node set N, and assume G is connected. Let P be any topology control protocol that preserves worst-case connectivity and let $G_P = (N, E_P)$ be the topology generated by P. Then, there exists a node placement such that the maximum physical node degree in G_P is $n - 1$, where $n = |N|$.*

Proof. Consider the node placement represented in Figure 9.2: there is one faraway node (node v) and all the other nodes are very close to each other. Among these nodes, node u is the closest to node v. Since the maxpower graph G is connected, node u is within

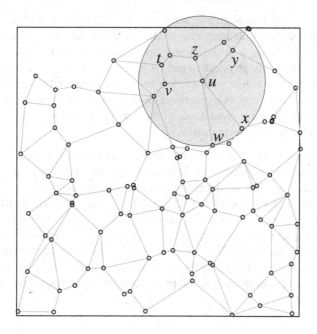

Figure 9.1 Difference between logical and physical node degrees: node u has logical degree equal to 5 and physical degree equal to 10.

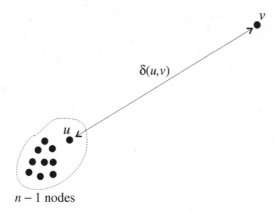

Figure 9.2 Example of node placement generating a topology in which at least one node has physical node degree equal to $n - 1$.

v's maximum transmitting range and, given the standard assumptions of symmetric wireless medium and that all the nodes have the same maximum transmit power, node v also is within u's maximum range. Since the nodes in the cluster are much closer to u than node v, and v is a neighbor of u in G, it follows that the physical degree of node u in the maxpower graph is $n - 1$. Let us now consider the topology $G_P = (N, E_P)$ generated by a worst-case connectivity preserving topology control protocol P. Can we do any better than having at least one node with physical degree equal to $n-1$? The answer is clearly *no*: since G_P must be strongly connected, at least one of the nodes in the cluster must have v within its transmitting range. Node u is the cluster node that is closer to node v: if node u has a connection to node v, its physical degree is $n-1$ for the reasons described above. If another cluster node $w \neq u$ has a connection to v, w's range is larger than $\delta(u, v)$, since u is the closest node to v, and every network node is within w's range. It follows that, in order to preserve connectivity, at least one of the network nodes must have physical degree equal to $n-1$, and the theorem is proven.

In other words, Theorem 9.3.3 states that if we want to ensure connectivity in the worst case, we must admit the possibility of having a 'high interference node', that is, a node whose communications affect all the remaining nodes in the network.

Another implication of Theorem 9.3.3 is that the gap between logical and physical node degrees can be very high in the worst case. In fact, as we will see in the following chapters, there exist topology control protocols that preserve worst-case connectivity and generate topologies with logical node degree bounded by a small constant (typically, 6). By Theorem 9.3.3, the physical node degree in these topologies can be as high as $n-1$, implying that the gap between logical and physical node degrees in a communication graph can be as high as $O(n)$.

10

Location-based Topology Control

In this chapter, we present two location-based distributed topology control protocols, the R&M protocol introduced in (Rodoplu and Meng 1999) and the LMST protocol presented in (Li et al. 2003).

As discussed in the previous chapter, this type of protocols relies on very accurate information (node locations), which is assumed to be somehow available to the nodes. The easiest way to satisfy this assumption is to equip every node in the network with low-power GPS receivers, which enable very accurate position estimation and loose synchronization with no message exchange between network nodes. This solution has a high cost in terms of hardware required on the nodes, but it entails no message overhead. An alternative solution is to use one of the many location estimation techniques introduced in the literature (see, for instance, (Niculescu and Nath 2003; Priyantha et al. 2000; Savvides et al. 2001)). These techniques usually assume that a subset of the network nodes (called the anchor nodes) is equipped with GPS receivers, and that nonanchor nodes can estimate their position by exchanging messages with the surrounding anchor nodes. In location estimation techniques, the reduced hardware cost due to the fact that only a subset of the nodes is GPS-equipped is traded off with the message overhead incurred to estimate node positions.

Before presenting the protocols, we remark that location-based topology control can be used in outdoor applications, while its applicability in indoor environments is limited by the poor accuracy provided by location estimation techniques in this setting.

10.1 The R&M Protocol

The R&M protocol (Rodoplu and Meng 1999) builds a topology that is optimized for the all-to-one communication pattern, where one of the network nodes is designated as the master node, and all the other nodes send messages to the master. This traffic pattern is typical of WSNs, where the deployed sensors must send the collected data to one (or more) base station(s). Given the focus on the all-to-one traffic pattern, and the fact that WSNs are typically deployed outdoor, the R&M protocol is well suited to build and maintain the topology in sensor networks.

Topology Control in Wireless Ad Hoc and Sensor Networks   P. Santi
© 2005 John Wiley & Sons, Ltd

Before presenting the protocol, we formally define the minimum-energy all-to-one communication problem.

Definition 10.1.1 (MinEnergyAllToOne problem) *Let $G = (N, E)$ be the maxpower graph (i.e. the communication graph obtained when all the nodes transmit at maximum power) and let $u \in N$ be the master node. For any given reverse spanning tree T of G rooted at u, the energy cost of T is defined as follows:*

$$ec(T) = \sum_{v \in N, v \neq u} c(v, p_T(v)),$$

where $c(v, p_T(v))$ is the energy cost of the link connecting node v to its parent $p_T(v)$ in T. The MinEnergyAllToOne problem is the problem of finding the reverse spanning tree of minimum energy cost rooted at the master node.

The R&M protocol is based on the notions of *relay region* and *enclosure graph*. Before introducing these notions, we need to define the power consumption model used to derive the link cost.

10.1.1   The power consumption model

As briefly discussed in Section 2.1, the propagation of the radio signal in the air can be modeled by the following components:

- the *path loss*, which models the average signal power received at a certain distance from the transmitter;

- *large-scale variations*, which model the randomness of the received signal strength over large distances;

- *small-scale variations*, which model the randomness of the received signal strength over small distances.

Since small-scale variations of the radio signal can be handled by diversity techniques and combiners at the physical layer, the power consumption model used in (Rodoplu and Meng 1999) accounts only for path loss and large-scale variations. Indeed, in order to simplify the presentation of the protocol, we consider only path loss. In fact, Rodoplu and Meng show that a minimum-energy design that addresses the increase in transmit power to handle large-scale variations is fundamentally the same as a design that accounts only for path loss.

Consider three nodes u, v, w positioned as in Figure 10.1. The main observation on which the R&M protocol is based is the classical 'triangle inequality' argument. Suppose node u wants to send a message to node v. In case of direct transmission, the minimum power needed to send the message is

$$P_{u \to v} = \delta(u, v)^\alpha,$$

where α is the path loss exponent. If node w is used as a relay, the total power dissipation equals

$$P_{u \to w \to v} = \delta(u, w)^\alpha + \delta(w, v)^\alpha + c_r,$$

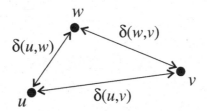

Figure 10.1 Depending on the relative distances between nodes u, v, and w, communicating using node w as a relay might be more energy efficient than direct communication from u to v.

where c_r is a constant term that accounts for the receiver power consumed at the relay node w. Since $\alpha \geq 2$ in outdoor environments, the using of the relay node might be more energy efficient for certain values of $\delta(u, v)$, $\delta(u, w)$, $\delta(w, v)$, and c_r. The goal of the R&M protocol is to identify such values, and to shape the network topology accordingly.

10.1.2 Relay region and enclosure graph

The relay region of a certain transmitter–relay node pair (u, w) identifies the set of points in the plane (node locations) for which communicating through the relay node is more energy efficient than direct communication. Formally,

Definition 10.1.2 (Relay region) *The relay region of a certain transmitter–relay node pair* (u, w) *is defined as*

$$RR_{u \to w} = \{(x, y) \in \mathbb{R}^2 : P_{u \to w \to (x,y)} < P_{u \to (x,y)}\}.$$

The shape of the relay region depends on the radio signal propagation model, and more in particular on the value of the distance-power gradient. The shape of the relay region for the cases $\alpha = 2$ and $\alpha = 4$ is reported in Figure 10.2.

The notion of relay region is used to define the *enclosure* of a node u, which represents the region of the plane beyond which it is not energy efficient for node u to search for one-hop neighbors. The formal definition of enclosure is quite complicated, but the intuition behind it is well illustrated in Figure 10.3.

Definition 10.1.3 (Enclosure and neighbor set) *Suppose the network nodes are deployed in a certain bounded two-dimensional region R. The enclosure of node u is defined as the set of points defined by the nonempty solution* $(\varepsilon_u, N(u))$ *to equations*

$$\varepsilon_u = \bigcap_{v \in N(u)} \left(RR^c_{u \to v} \cap R \right)$$

and

$$N(u) = \{v \in N | (x_v, y_v) \in \varepsilon_u, v \neq u\},$$

where $RR^c_{u \to v}$ *is the complement of set* $RR_{u \to v}$ *and* (x_v, y_v) *are the coordinates of node v. $N(u)$ is called the neighbor set of node u, and every element in $N(u)$ is said to be a neighbor of u.*

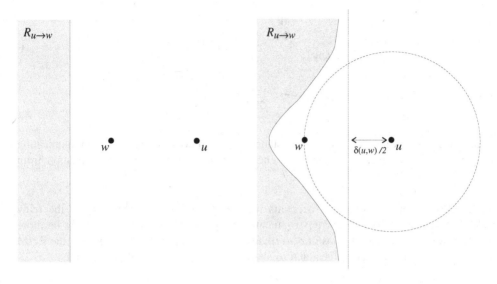

Figure 10.2 Relay region of the (u, w) node pair when the distance-power gradient is $\alpha = 2$ (left) and $\alpha = 4$ (right).

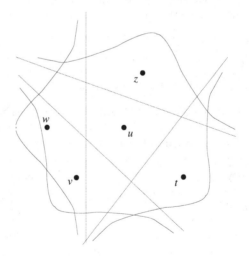

Figure 10.3 Enclosure of node u. Nodes t, v, w, z are the neighbors of u in the enclosure graph.

In (Rodoplu and Meng 1999) it is shown that the pair $(\varepsilon_u, N(u))$ always exists and is unique, that is, that the definition of enclosure is mathematically sound.

Definition 10.1.4 (Enclosure graph) *Let N be a set of nodes deployed in a certain bounded two-dimensional region R. The enclosure graph of N is the directed graph $G_\varepsilon = (N, E_\varepsilon)$, where the directed edge $(u, v) \in E_\varepsilon$ if and only if $v \in N(u)$.*

The enclosure graph enjoys several nice properties (see (Rodoplu and Meng 1999)):

- it preserves connectivity in the worst case: if the maxpower graph on node set N is strongly connected, then the enclosure graph $G_\varepsilon = (N, E_\varepsilon)$ is strongly connected as well;

- it is sparse;

- it contains the minimum-energy reverse spanning tree rooted at any master node $u \in N$;

- it can be computed in a fully distributed and localized fashion.

10.1.3 Protocol description

The R&M protocol for finding the optimal solution to the MINENERGYALLTOONE problem is composed of two phases: first, every node in the network computes its enclosure and neighbor set, that is, its local view of the enclosure graph G_ε; then, a global cost distribution phase is initiated to determine the minimum-energy reverse spanning tree rooted at the master node, which is a subgraph of G_ε.

Phase 1: Compute the enclosure graph.   We present the algorithm for a generic node u. Initially, node u broadcasts a beacon message containing its ID and position at maximum transmit power. As beacon messages of other nodes are received, u calculates the corresponding relay region, and keeps track of whether the newly found node is in the relay region of some previously found node. To accomplish this task, potential neighbors are marked *dead* or *alive*: a node is said to be dead if it lays in the relay region of some other neighbor of u, and it is said to be alive otherwise. After all the beacon messages from surrounding nodes are received, the set of alive nodes defines the neighbor set of node u, that is, u's local view of the enclosure graph.

The Phase 1 of the protocol uses an auxiliary function called *FlipAllStatesDownChain* to update the alive/dead state of the currently discovered nodes. This function is needed to handle the following situation: assume node v is marked dead, because it is in the relay region of node w (we say that node w *blocks* node v). Later on, node u receives the beacon sent by a third node z, which blocks node w but not node v. In this case, node w is marked dead, and node v should be 'revived' since it is no longer blocked by any node. Indeed, there may be a chain of nodes in which one node blocks the next one in the chain; when a newly found node blocks the first alive node in such a chain, the states of all the nodes down in the chain need to be flipped.

The protocol for computing the enclosure graph is reported in Figure 10.4, and the auxiliary function *FlipAllStatesDownChain* is reported in Figure 10.5.

Phase 2: Cost distribution.   In the second phase of the protocol, the classical distributed Bellman–Ford shortest path algorithm (Lynch 1996) is used on the enclosure graph to compute the minimum-energy reverse spanning tree rooted at the master node. The cost metric used to compute the shortest path between any node and the master is the energy cost of the link, as predicted by the path loss model. Each node calculates the minimum

Algorithm ENCLOSUREGRAPH:
(algorithm for node u)

AliveNodes is the current set of alive nodes
DeadNodes is the current set of dead nodes
$N(u)$ is the neighbor set of node u
ε_u is the enclosure of node u

1. Initialization
1.1 $AliveNodes = \emptyset$
1.2 $DeadNodes = \emptyset$
1.3 $N(u) = \emptyset$
1.4   send beacon message $(u, (x_u, y_u))$

2. Upon receiving beacon message $(v, (x_v, y_v))$
2.1   compute relay region $RR_{u \rightarrow v}$
2.2 $MarkDead(v)$
2.3 $FlipAllStatesDownChain(v)$

3. Termination (after receiving all beacon messages)
3.1 $N(u) = AliveNodes$
3.2 $\varepsilon_u = \bigcap_{v \in AliveNodes} \left(RR^c_{u \rightarrow v} \cap R \right)$

Figure 10.4   Algorithm for constructing the enclosure graph.

Function *FlipAllStatesDownChain(v)*:
(algorithm for node u)

1. If $v \in AliveNodes$ then
      $MarkDead(v)$
      for each $w \in RR_{u \rightarrow v}$ do $FlipAllStatesDownChain(w)$

   else
      if $(v \notin RR_{u \rightarrow w}) \; \forall w \in AliveNodes$ then
      $MarkAlive(v)$
      for each $w \in RR_{u \rightarrow v}$ do $FlipAllStatesDownChain(w)$

Figure 10.5   Auxiliary function *FlipAllStatesDownChain*.

cost to reach the master node given the cost of its neighbors: when node u receives the information $Cost(v)$ from node $v \in N(u)$, it computes the cost of the path to the master node passing from v as follows:

$$C_{u,v} = Cost(v) + \delta(u, v)^\alpha + c_r.$$

Then, node u computes the minimum energy cost of sending a message to the master node as

$$Cost(u) = \min_{v \in N(u)} C_{u,v},$$

and broadcasts the message $Cost(u)$ at maximum power. The calculation of $Cost(u)$ is repeated each time a new cost message is received, and a new cost message is broadcast by u each time $Cost(u)$ is updated. After a finite number of iterations, the algorithm stabilizes (see (Lynch 1996)). The neighbor of u that is the next hop in the minimum-energy path to the master node is the parent of u in the energy-optimal reverse spanning tree T_{opt}. After all the nodes in the network have computed their cost and thus determined their minimum cost neighbor link, the construction of the optimal topology T_{opt} is complete. Note that T_{opt} is built in a top-down fashion: the first links that stabilize are those connecting the master node with its immediate neighbors; then, stabilization propagates downward in the reverse tree.

10.1.4   Discussion

The R&M protocol presented in this section has the nice feature of building a topology that optimizes a certain energy-efficient topology problem, namely, MINENERGYALLTOONE. However, the computation of the optimal topology requires global information exchange (Phase 2 of the protocol), causing a certain message overhead.

Given its features (building an energy-efficient topology for communicating with a master node, but with rather high message overhead), the R&M protocol can be successfully used in many WSN application scenarios, where mostly stationary network nodes must send their data to a gateway node. In this case, the network topology is typically built at the beginning of the network lifetime, and constructing an energy-efficient topology (even if this comes at the expense of a relatively high message overhead) is a primary design concern.

It is interesting to note that the reverse version of MINENERGYALLTOONE, that is, the energy-efficient broadcast problem, is NP-hard. The reason for this gap in computational complexity is due to the fact that, in the energy-efficient broadcast problem, the nodes use the so-called wireless advantage, that is, the fact that the message sent by a node can be received by all the nodes within its range, and computing the best way of using the 'wireless advantage' is NP-hard. On the other hand, in the MINENERGYALLTOONE problem, the goal is to find the most efficient way of communicating with the master node, that is, the 'wireless advantage' is not useful in this case. Then, to solve the problem, it is sufficient to find the most energy-efficient path from any node to the master node, which can be done in polynomial time.

A final observation concerns one of the potential disadvantages of the R&M protocol, that is, the fact that it relies on an explicit radio signal propagation model that is used to compute the enclosure graph and T_{opt}. A consequence of this fact is that the topology generated by R&M might be different from the optimal one if the actual channel conditions are different from those predicted by the channel model used to compute T_{opt}.

10.2   The LMST Protocol

In Chapters 7 and 8, we have seen that the MST enjoys several nice features: it preserves connectivity, it is extremely sparse, and it is a h-approximation (for some constant $h > 0$) of the optimal solution to the RA, to the WSRA, and to the energy-efficient
broadcast problem. The drawback of this topology is that its computation requires the
exchange of global information. To circumvent this problem, Li et al. introduced the
LMST (Local MST) protocol (Li et al. 2003), which computes a local approximation of
the MST.

10.2.1   Protocol description

The LMST protocol is composed of three phases: *information exchange, topology construction*, and *determination of transmit power*, and an optional optimization phase: *construction
of topology with bidirectional links*.

   In the protocol specification, it is assumed that all the nodes have the same maximum
transmit power and that the wireless medium is symmetric.[1] We also adopt the conventional
definition of maxpower graph, that is, the graph $G = (N, E)$ obtained when all the nodes
transmit at maximum power.

Definition 10.2.1 (Visible neighbors) *The visible neighbors of node $u \in N$ are all the one-
hop neighbors of u in the maxpower graph $G = (N, E)$. Formally,*

$$VN_u = \{v \in N | (u, v) \in E\} \cup \{u\}.$$

Information exchange.   In the first phase of the protocol, each node sends its ID and
location to all nodes in the visible neighborhood. This can be accomplished by sending a
beacon message at maximum transmit power.

Topology construction.   Once all the beacon messages of the visible neighbors have been
received, each node constructs its local MST by applying the classical Prim's algorithm
(Prim 1957). The link weight used to build the MST is its length (Euclidean distance).
This choice is compatible with any path loss model, as the power cost of link (u, v) is
proportional to $\delta(u, v)^\alpha$, with $\alpha \geq 1$. So, the MST that results using any path loss model as
the weight function is the same one that is built using the Euclidean distance.

   Note that the MST produced by Prim's algorithm might not be unique. Since uniqueness
of the MST is needed to prove that LMST preserves connectivity (see next subsection), the
authors of (Li et al. 2003) define a link-weight function that also accounts for the lexical
order of the node Ids of the link endpoints.

   After Prim's algorithm execution, every node u in the network knows the (unique) MST
$T_u = (VN_u, E_u)$ connecting u to all its visible neighbors. The next step is to define the set
of neighbors of u in the final topology, that is, u's local view of the LMST topology. This
is accomplished by defining the neighbor relation as follows:

---

[1]The wireless medium is symmetric if the fact that node u can reach node v using a certain power P implies
that node v also can reach node u using power P.

Definition 10.2.2 (Neighbor relation) *Node v is a neighbor of node u, denoted as $u \rightarrow v$, if and only if v is a one-hop neighbor of u in its minimum spanning tree $T_u = (VN_u, E_u)$. Formally,*

$$u \rightarrow v \Leftrightarrow (u, v) \in E_u.$$

The neighbor set of node u, denoted as $N(u)$, is defined as

$$N(u) = \{v \in VN_u | u \rightarrow v\}.$$

The topology produced by LMST is obtained by connecting each node to its neighbors:

Definition 10.2.3 (LMST topology) *The final topology generated by the LMST protocol is the directed graph $G_{LMST} = (N, E_{LMST})$, where directed edge $(u, v) \in E_{LMST}$ if and only if $u \rightarrow v$.*

Note that G_{LMST} is not a simple superposition of all the local MSTs, since only immediate neighbors in the local MSTs are included in G_{LMST}.

Determination of transmit power. The final step of the protocol is the determination of the transmit power needed to send a message to any neighbor node. This can be accomplished by measuring the received power of the beacon messages: when node u receives the beacon from a certain visible neighbor v, it can estimate the minimum power level needed to reach v by comparing the received power of the beacon with the maximum transmit power (we recall that all beacons are sent at maximum power). This approach can be applied in combination with any propagation model.

The nodes also compute the *broadcast power*, that is, the minimum power needed to reach the farthest node in $N(u)$.

Construction of topology with bidirectional links (optional). As shown in Figure 10.6, the topology built by LMST might contain unidirectional links. Since this type of links should be avoided in the final topology (see Section 7.4 and Chapter 9 for a discussion

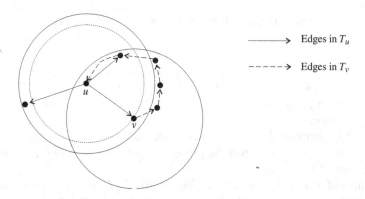

Figure 10.6 Example showing that the G_{LMST} topology may contain unidirectional links. Solid circles represent u's and v's maximum transmitting range.

of this issue), the authors of (Li et al. 2003) propose two techniques for avoiding this inconvenience: (i) enforcing all unidirectional links in G_{LMST} to become bidirectional; or (ii) deleting all the unidirectional links in G_{LMST}. If technique (i) is used, the obtained graph is the symmetric supergraph of the original topology, which is denoted as G_{LMST}^{+}; if technique (ii) is used, we obtain the symmetric subgraph of the original topology, which we call G_{LMST}^{-}.

Note that which one of the G_{LMST}^{+} and the G_{LMST}^{-} topology is to be preferred depends on the application scenario: the former topology is relatively dense, and it thus keeps more routing redundancy, which is useful to balance the traffic load and to improve fault tolerance; the latter topology is very sparse, and it should be used when the expected network traffic is quite low.

To convert G_{LMST} into either G_{LMST}^{+} or G_{LMST}^{-}, each node u after Phase 3 of the algorithm probes all the nodes in $N(u)$ to find out whether the corresponding link is unidirectional. In case it is, the link is either removed (technique (ii)), or the neighbor node is notified to add the reverse edge (technique (i)). Note that in case the probed link (u, v) is unidirectional node u is not in $N(v)$, so node v does not know which transmit power to use to the send the reply message to u. This problem can be easily solved by piggybacking the transmit power mp_v used by u to send the probe message in the message itself. Given the assumption of symmetric wireless medium, using power mp_v is sufficient for node v to reach node u.

Protocol LMST is summarized in Figure 10.7.

10.2.2   Protocol analysis

In (Li et al. 2003) it is proven that the topology produced by LMST has the following properties:

1. it preserves connectivity in the worst case;

2. it has maximal logical node degree equal to 6;

3. it can be computed in a fully distributed and localized fashion; in particular, computing G_{LMST} requires sending only n messages overall (n is the number of network nodes).

Properties (1) and (2) are satisfied also by the symmetric variants of G_{LMST}, G_{LMST}^{+} and G_{LMST}^{-}. As for the message complexity, computing both G_{LMST}^{+} and G_{LMST}^{-} requires exchanging $O(n^2)$ messages overall (at most $n-1$ probe messages are sent by a node in Phase 4 of the protocol).

Note that the upper bound stated in (2) is on the number of *logical* neighbors; it is easy to find worst-case scenarios in which the physical degree of a node in G_{LMST} is arbitrarily high (this is implied by Theorem 9.3.3).

The authors of (Li et al. 2003) have also evaluated the average-case performance of LMST on random node deployments through simulation, and they have verified that LMST produces topologies with a smaller average logical node degree and average transmission radius with respect to those generated by R&M and CBTC.

Algorithm LMST:
(algorithm for node u)

VN_u is the visible neighborhood of node u
$N(u)$ is the neighbor set of node u
bp_u is the broadcast power of node u
(x_u, y_u) are the coordinates of node u

1. Information exchange
   send beacon $(u, (x_u, y_u))$ at maximum power
   upon receiving beacon $(v, (x_v, y_v))$, store the received power of this message in rp_v

2. Topology construction (after all beacons have been received)
   build the local MST on nodes in VN_u using Prim's algorithm
   let $T_u = (VN_u, E_u)$ be this local MST
   $N(u) = \{v \in VN_u | (u, v) \in E_u\}$

3. Determination of transmit power
   for each $v \in N(u)$
      compute the minimum power mp_v needed to reach v based on rp_v
   $bp_u = \max_{v \in N(u)} mp_v$

4. (Optional) Topology with bidirectional links
   for each $v \in N(u)$
      send probe message (u, mp_v) to v using power mp_v
      upon receiving reply message $(v, state)$ from v
         if $state = uni$ then
            notify v sending message (u, add) (technique 1)) using power mp_v
         or
            delete v from $N(u)$ (technique (2))
   upon receiving probe message (v, mp_u)
      if $v \in N(u)$ then
         send reply message (u, bi) using power mp_u
      otherwise
         send reply message (u, uni) using power mp_u
   upon receiving notify message (v, add)
      add node v to $N(u)$, with associated power mp_v
      if necessary, update the broadcast power bp_u

Figure 10.7 The LMST protocol.

10.2.3   The FLSS$_k$ protocol

Some of the authors of (Li et al. 2003) have presented a variation of the LMST algorithm aimed at improving the fault tolerance of the constructed topology. In particular, the design goal is to build an energy-efficient topology that preserves k-connectivity (provided the maxpower communication graph is k-connected), where k is a small constant (typically, 2–3). The resulting protocol, presented in (Li and Hou 2004), is called FLSS$_k$ (Fault-tolerant Local Spanning Subgraph).

Similarly to LMST, FLSS$_k$ is composed of three phases: information exchange, topology construction, and determination of transmit power. The information exchange phase is identical to that of LMST: every node broadcasts its ID and position at maximum power, and collects the location information sent by its visible neighbors. The main difference between LMST and FLSS$_k$ is in the topology construction phase: instead of building a local MST on the set of its visible neighbors, a node u builds a spanning subgraph G'_u that preserves k-connectivity on the same set of nodes (see (Li and Hou 2004) for details). Then, node u selects its immediate neighbors in the G'_u graph as logical neighbors that are retained in the final topology. The last phase of the protocol (determination of transmit power) is the same as in LMST.

Similar to LMST, the topology built by FLSS$_k$ might contain unidirectional links, and symmetry can be enforced by either removing all the unidirectional links or by making them bidirectional.

Li and Hou prove that FLSS$_k$ (and its symmetric variants) preserves k-connectivity, and that it minimizes the maximum transmitting range of nodes in the network over all localized algorithms. Furthermore, Li and Hou investigate the average-case performance of FLSS$_k$ through simulation, whose results show that FLSS$_k$ is more energy efficient than other existing localized fault-tolerant topology control protocols, such as the k-UPVCS algorithm introduced in (Hajiaghayi et al. 2003) and the k-connected variation of CBTC introduced in (Bahramgiri et al. 2002).

11

Direction-based Topology Control

In this chapter, we consider topology control protocols that rely on the ability of the nodes to estimate the relative direction of their neighbors. This is relatively less accurate information than knowing exact node locations, as the former type of information can be determined if the latter is known, but not vice versa.

Several techniques for estimating the direction from which a certain node is transmitting have been proposed and discussed in the IEEE Antenna and Propagation community (IEEE 2004). This problem is known as the Angle-of-Arrival (AoA) problem, and it is typically solved by equipping nodes with more than one directional antenna (Krizman et al. 1997). So, in the case of directional information also, some extra hardware on the nodes (with respect to the standard assumption of nodes equipped with a single, omnidirectional antenna) is needed in order to provide the requested information. An advantage of using AoA-based techniques instead of location-based techniques is that the AoA can be accurately estimated in indoor environments also.

Despite the relatively less accurate information used, direction-based topology control protocols can produce almost as good topologies as in the case of location-based topology control. In particular, fully distributed, localized protocols that preserve worst-case connectivity can be designed in this setting also.

In the remainder of this chapter, we present two location-based topology control protocols: the CBTC protocol introduced in (Wattenhofer et al. 2001) and further analyzed in (Li et al. 2001) and the DISTRNG protocol presented in (Borbash and Jennings 2002).

11.1 The CBTC Protocol

The CBTC (Cone-based Topology Control) protocol (Li et al. 2001; Wattenhofer et al. 2001) is based on the following idea: set the transmit power level of node u to the minimum value $p_{u,\rho}$ such that u can reach at least one node in every cone of width ρ centered at u (see Figure 11.1). In other words, a node must retain connections to at least one neighbor in 'every direction', where parameter ρ determines the granularity of what is meant by 'every direction'.

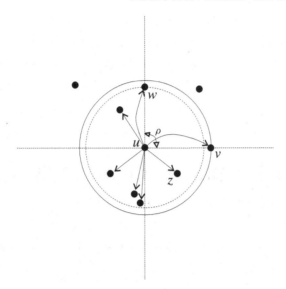

Figure 11.1 Intuition behind the CBTC protocol: node u sets its power level to the minimum value $p_{u,\rho}$ such that it can reach at least one node in every cone of width ρ centered at itself. In the example above, $\rho = \frac{\pi}{2}$, and node u must use a transmit power level at least sufficient to reach node v. If a lower power is used, the angular gap between u's neighbors would be $> \frac{\pi}{2}$ (see nodes w and z), and the condition that every cone of width $\frac{\pi}{2}$ centered at u contains at least one neighbor would be violated.

Note that the idea behind CBTC is very similar to that used in the definition of the Yao Graph (see Appendix A). We recall that the Yao Graph of parameter c (for some integer $c \geq 6$), denoted YG_c, is defined as follows: at each node u, divide the plane into c equally separated cones centered at u; then, connect u to its closest neighbor within each cone. The difference between YG_c and the topology generated by CBTC is depicted in Figure 11.2. In order to make the two graphs as much similar as possible, we set $c = 6$ and $\rho = \frac{\pi}{3}$. In YG_6, the cones are predefined, and it is sufficient to reach one neighbor in each such cone. On the contrary, in CBTC, it is required that the angular gap between any two neighbors of u is at most ρ; in other words, when a cone of width ρ centered at u sweeps the plane, it must always contain at least one neighbor. This is a stronger requirement than in case of YG_6, as shown in Figure 11.2.

We now present the distributed implementation of CBTC, and then we discuss its properties. Finally, we describe some of the variations of CBTC that have appeared in the literature.

11.1.1 The basic CBTC protocol

The CBTC protocol is composed of two phases: in the first phase (basic protocol), every node u determines the minimum power $p_{u,\rho}$ needed to reach a neighbor in 'every direction' as described above; then, network nodes exchange additional information to identify energy-inefficient edges, which are removed from the final topology.

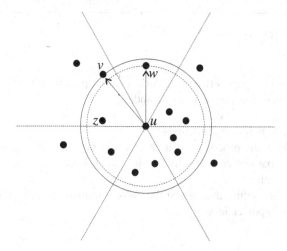

Figure 11.2 Difference between YG_6 and the topology generated by *CBTC* with parameter $\rho = \frac{\pi}{3}$. The dotted lines define the six cones used in the definition of YG_6; node u must use sufficient transmit power to reach the closest neighbor in each cone, which corresponds to the power needed to reach node w (dotted circle). However, using this transmit power is not sufficient to fulfill CBTC's condition: there exists a cone of width $\frac{\pi}{3}$ that contains no node (see the angular gap between nodes w and z). To fill this gap, u's transmit power level must be increased to reach node v also (solid circle).

CBTC uses two types of messages: *beacon* messages, which are sent at a certain power $p \le P_{\max}$ ($P_{\max}$ denotes the maximum nodes' transmit power, which is assumed to be the same for all the nodes) and received by all the nodes that are within u's range with power p; and *acknowledgment* messages (Ack for short), which are sent in response to beacons and are received only by the node that originated the beacon.

The beacon message contains the node ID and the power p used to send the message; the Ack message contains the ID of the sender, the ID of the intended receiver (the node that originated the beacon), and the power used to transmit the message. Inclusion of the transmit power in the messages is needed to identify energy-inefficient links in Phase 2 of the protocol.

The first phase of CBTC is as follows. Initially, node u sends the beacon at power p_0 and collects the Ack messages sent by the nodes that received the beacon. When receiving an Ack message, node u stores the identity of the new neighbor and determines its relative direction. As discussed at the beginning of this chapter, this is made possible by the use of AoA estimation techniques, such as using multiple directional antennas. The Ack messages are sent using the same power level used to send the originating beacon message. This way, under the common assumption of symmetric wireless medium, we can ensure that node u eventually receives the acknowledgments from all the nodes that received its beacon. After all the Acks for power level p_0 have been collected, node u invokes the *CheckGap* procedure, which verifies whether the condition on the angular gap between neighbors is met. If the condition is not satisfied, node u invokes the procedure *IncPower*, which increases the transmit power level to the next level p_1. Then, it sends a new beacon message, waits

Algorithm BASICCBTC:
(algorithm for node u)

ρ is the required angular gap between neighbors (input parameter)
$p(u)$ is the current transmit power level of node u
$N(u)$ is the neighbor set of node u
$D(u)$ is the set of u's neighbor directions
$CheckGap(\rho, D(u))$ is the procedure that checks whether the CBTC condition with
                 parameter ρ is satisfied. It returns **True** if it is satisfied, **False**
                 otherwise
$IncPower(p)$ is the procedure that, given the current transmit power p, returns the next
                 transmit power level

1. Initialization
    $N(u) = \emptyset$
    $D(u) = \emptyset$
    $p(u) = 0$

2a. Computing power $p_{u,\rho}$
    repeat until $CheckGap(\rho, D(u)) =$ **True** or $(p(u) = P_{\max})$
        $p(u) = IncPower(p(u))$
        send beacon $(u, p(u))$ at power $p(u)$
        repeat until all Acks have been received
            receive Ack $(v, u, p(v))$ from node v
            $N(u) = N(u) \cup \{v\}$
            update direction set $D(u)$ including v's direction

2b. Sending Ack message
    upon receiving beacon $(v, p(v))$ from v
        check if this is the first beacon received from v
        if *yes* send Ack $(u, v, p(v))$ at power $p(v)$, otherwise ignore the beacon

3. Finalization
    $p_{u,\rho} = p(u)$

Figure 11.3 The BASICCBTC protocol.

for the Acks, and so on. This algorithm is repeated until either the condition on the angular
gap between neighbors is satisfied or $p_i = P_{\max}$. Phase 1 of CBTC (also called BASICCBTC)
is summarized in Figure 11.3.

Note that the following optimization, called the *shrink back* operation, can be easily
implemented. At the end of BASICCBTC's execution, a node sets its transmit power at the

maximum level if the condition on cone coverage cannot be satisfied. We call such nodes *boundary nodes*. The shrink back operation is executed at boundary nodes only, with the purpose of reducing the broadcast power $p_{u,\rho}$ without reducing the cone coverage. More specifically, BASICCBTC is modified in such a way that, at each iteration, a node in $N(u)$ is tagged with the power used the first time it was discovered. Suppose the power levels used during the neighbor discovery phase are $p_0, p_1, \ldots, p_k = P_{\max}$, and let CC_i be the cone coverage provided by the neighbors at level i. If $CC_k < 2\pi$, the broadcast power level $p_{u,\rho}$ is reduced to the minimum level p_i such that $CC_i = CC_k$. Note that tagging each neighbor with the minimum power needed to reach it is useful for implementing another optimization also: if u must send a packet to a certain neighbor v that can be reached with power $p_i < p_{u,\rho}$, it can send the packet using power p_i instead of the broadcast power.

11.1.2   Dealing with asymmetric links

Let us denote with G_{CBTC}^{ρ} the graph obtained after BASICCBTC's execution with parameter ρ, that is, the graph that contains the directed link (u, v) if and only if $v \in N(u)$ at the end of the protocol. The example reported in Figure 11.4 shows that the neighbor relation induced by BASICCBTC is not symmetric, that is, G_{CBTC}^{ρ} can contain unidirectional links. Suppose ρ is set to $\frac{2}{3}\pi$. At the end of BASICCBTC's execution, node u sets its transmit power to the minimum level $p_{u,\frac{2}{3}\pi}$ needed to reach the three neighbors at distance d. Since the distance between u and v is greater than d, there is no direct link between u and v. On the other hand, node v has u in its neighbor list at the end of the protocol. In fact, if v would use a lower power than the minimum power $p_{v,\frac{2}{3}\pi}$ needed to reach u, the angular gap between its neighbors would be greater than $\frac{2}{3}\pi$ (see the gap between nodes w and z). So, the directed link (v, u) is part of the final topology, while the reverse link is not.

Since, as discussed in several parts of this book, having a topology with symmetric links is desirable, the authors of (Li et al. 2001; Wattenhofer et al. 2001) propose two techniques to address unidirectional connections: (i) augmentation and (ii) asymmetric edge removal.

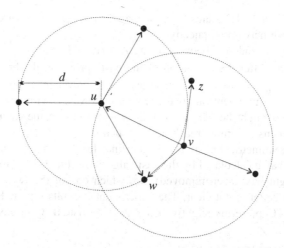

Figure 11.4   Example of asymmetric link with BASICCBTC. The parameter ρ is set to $\frac{2}{3}\pi$.

In (i), every asymmetric link (u, v) is made symmetric by adding the reverse edge (v, u) in the graph.[1] To implement this strategy, it is sufficient that every node u at the end of BASICCBTC advertises its neighbor set at the broadcast power $p_{u,\rho}$. Upon receiving the neighbor set from v, node u verifies whether $v \in N(u)$; if yes, no action is taken; otherwise, v is included in $N(u)$, and u's broadcast power is increased consequently (if necessary). In the following, we will call this version of CBTC as AUGMCBTC, and we will denote the corresponding topology with $G^{\rho,+}_{\text{CBTC}}$.

In (ii), asymmetric links are removed from the final topology as follows.[2] After finishing BASICCBTC, a node u sends a message to each node $v \notin N(u)$ to which it sent an Ack, telling v to remove u from $N(v)$. As a consequence of this action, the broadcast power $p_{v,\rho}$ of node v might be reduced. In the following, we will call this version of CBTC as REMCBTC, and we will denote the corresponding topology with $G^{\rho,-}_{\text{CBTC}}$.

11.1.3 Protocol analysis

The following theorems have been proven in (Li et al. 2001).

Theorem 11.1.1 (Li et al. 2001) *Let G be the maxpower communication graph, and assume G is connected. Let $G^{\rho,+}_{\text{CBTC}}$ be the topology generated by AUGMCBTC. Then $G^{\rho,+}_{\text{CBTC}}$ is (worst-case) connected if and only if $\rho \leq \frac{5}{6}\pi$.*

In words, Theorem 11.1.1 states that, if $\rho \leq \frac{5}{6}\pi$ and the maxpower graph G is connected, then the topology remains connected after AUGMCBTC's execution. On the other hand, if $\rho > \frac{5}{6}\pi$, there exists a node placement such that G is connected, but $G^{\rho,+}_{\text{CBTC}}$ is not connected. An example of such node placement is reported in (Li et al. 2001).

Theorem 11.1.2 (Li et al. 2001) *Let G be the maxpower communication graph, and assume G is connected. Let $G^{\rho,-}_{\text{CBTC}}$ be the topology generated by REMCBTC. If $\rho \leq \frac{2}{3}\pi$, then $G^{\rho,-}_{\text{CBTC}}$ is (worst-case) connected.*

In words, Theorem 11.1.2 states that, as long as $\rho \leq \frac{2}{3}\pi$, removing asymmetric links does not compromise network connectivity.

Note that there is a trade-off between using AUGMCBTC with $\rho = \frac{5}{6}\pi$ and using REMCBTC with $\rho = \frac{2}{3}\pi$. After the execution of the basic protocol, the broadcast power level of node u with $\rho = \frac{5}{6}\pi$ is lower than or equal to the power level with $\rho = \frac{2}{3}\pi$ (because of the less stringent requirement on cone coverage). However, with augmentation, the final level used by node u might be *increased* with respect to the value calculated by the basic algorithm (this happens if u must reach node v such that $u \in N(v)$ but $v \notin N(u)$). On the other hand, with asymmetric link removal, the final level used by u might be *decreased* with respect to the value calculated by the basic algorithm (this is because some of the links incident into u might have been removed). So, which one of the two symmetric versions of CBTC performs better is not clear. The experimental results reported in (Li et al. 2001) show that REMCBTC performs slightly better than AUGMCBTC in case of random node deployment.

---

[1]This corresponds to computing the symmetric supergraph of G^{ρ}_{CBTC}.
[2]This corresponds to computing the symmetric subgraph of G^{ρ}_{CBTC}.

11.1.4   Removing energy-inefficient links

A final optimization phase can be applied to both the symmetric versions of CBTC, with the purpose of further reducing the transmission power of each node. This optimization requires that nodes have the ability to perform some sort of distance estimation. In particular, for any pair v, w of u's neighbors, node u must be able to determine which one of them is closer. This can be accomplished by comparing the transmit powers included in the incoming messages received from v and w (we recall that this information is included in both beacon and Ack messages) with the reception powers of the messages.

The goal of this optimization stage is to identify energy-inefficient links, which can be removed without impairing network connectivity. These are called *redundant edges*, and are defined as follows:

Definition 11.1.3 (Redundant edge) *Let v, w be neighbors of u in the final topology, and assume that $\delta(v, w) < \max\{\delta(u, v), \delta(u, w)\}$. Then, the longer of the edges (u, v) and (u, w) is redundant.*

In (Li et al. 2001), it is shown that redundant edges can be removed from the final topology without impairing network connectivity. However, removing too many edges from the final topology might be a disadvantage because, for instance, the paths between nodes would become too long. Since CBTC's goal is to reduce the average transmit power of the nodes, the choice is then to remove only redundant edges with length greater than the longest nonredundant edge.

11.1.5   Discussion

The CBTC protocol enjoys several nice features: it is fully distributed, localized, preserves network connectivity, and requires only directional information. This explains the popularity of CBTC, which is probably the most famous topology control protocol. However, CBTC has a weak point, namely, the relatively high number of messages that must be exchanged to compute the network topology.

The reasons for this relatively high message overhead are three: (i) the beacon-Ack message exchange needed to estimate neighbor directions; (ii) the mechanism used to discover new neighbors, based on sending beacons with increasing transmit power; and (iii) the further message exchange needed to render the final topology symmetric. In particular, the choice of the power increase strategy in BASICCBTC (the *IncPower* procedure) is quite critical: on the one hand, starting with a very low transmit power p_0 and increasing the power level at each step by a small quantity ε might cause the sending of an excessive number of beacon messages; on the other hand, if the power levels used for beaconing are very few, then the number of new neighbors discovered at each step is high, resulting in computing of a very rough estimate of the broadcast power $p_{u,\rho}$. The choice of the better power increase strategy is scenario dependent: if the expected node density is very high, performing a very accurate neighbor discovery (i.e. using many different power levels for beaconing) is probably the right choice; on the contrary, if the expected density is low, using relatively few power levels for beaconing is preferable.

11.1.6   CBTC variants

Several variants of the CBTC protocol have been proposed recently. In particular, in (Bahramgiri et al. 2002), Bahramgiri et al. discuss the conditions under which CBTC ensures k-connectivity. They prove the following theorem:

Theorem 11.1.4 (Bahramgiri et al. 2002) *Let G be the maxpower communication graph, and assume G is k-connected, for some constant $k > 0$. Let $G_{CBTC}^{\rho,-}$ be the topology generated by REMCBTC. If $\rho \leq \frac{2\pi}{3k}$, then $G_{CBTC}^{\rho,-}$ is (worst-case) k-connected.*

Furthermore, they discuss necessary conditions for achieving k-connectivity with REM-CBTC. Finally, they introduce a three-dimensional version of CBTC, in which the notion of coverage is extended to three-dimensional cones.

Another variation of CBTC has been presented in (Huang et al. 2002), where Huang et al. introduce an implementation of CBTC based on the use of directional antennas. In fact, it must be noted that although CBTC requires the use of directional antennas to estimate AoA these antennas are not used to exchange messages: all the communications in the original CBTC protocol are performed using omnidirectional antennas. On the contrary, the protocol introduced in (Huang et al. 2002) makes explicit use of directional antennas: the plane is divided into sectors (corresponding to the possible orientations of the antenna), and the minimum powers needed to reach at least one neighbor in each sector are computed. Note that this is indeed a distributed computation of the Yao Graph (for a discussion of CBTC and Yao Graph see above).

11.2   The DISTRNG Protocol

The DISTRNG introduced in (Borbash and Jennings 2002) is a distributed implementation of the computation of the Relative Neighborhood Graph (RNG), which is defined as follows (see also Appendix A):

Definition 11.2.1 (Relative neighborhood graph) *Let N be a set of points in the Euclidean two-dimensional space. The Relative Neighborhood Graph of N, denoted by RNG(N), has an edge between two nodes u and v if there is no node $w \in N$ such that $\max\{\delta(u, w), \delta(v, w)\} \leq \delta(u, v)$.*

Let *Lune*(u, v) denote the intersection of the circles of radius $\delta(u, v)$ centered at u and v, respectively. Intuitively, edge (u, v) belongs to *RNG(N)* if and only if no other node in N lays in *Lune*(u, v) (see Figure 11.5). Note that, contrary to the case of CBTC, the neighbor relation in *RNG* is symmetric.

The *RNG* topology has several interesting features, as evidenced by the simulation-based investigation on random node deployments reported in (Borbash and Jennings 2002). The authors considers several aspects of the generated topology

- average logical node degree;

- hop diameter;

- maximum and average node transmitting range;

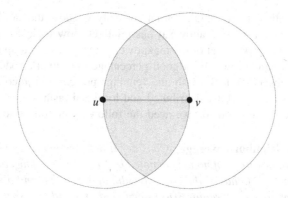

Figure 11.5 Definition of *RNG(N)*: edge (u, v) is in *RNG(N)* if and only if *Lune(u, v)* (shaded area) contains no other node in N.

- connectivity;

- size of the largest biconnected component.

Intuitively, a good topology is one in which nodes have low degree, the hop-diameter is close to the one of the maxpower communication graph, the nodes have small transmitting range, and connectivity (possibly, also biconnectivity) is ensured. Clearly, identifying a topology that satisfies all these goals at the same time is virtually impossible, as many of them are conflicting with each other. The objective is then to build a topology that is a good compromise between the above goals. To identify such topology, Borbash and Jennings performed extensive simulation on random node deployments, measuring the above listed parameters for the following topologies: the *MST*, the *RNG*, and the *minR* graph, which is obtained by finding the smallest common transmitting range such that connectivity is achieved (i.e. the CTR), and connecting the nodes consequently.

The simulation-based investigation has shown the following:

- *Logical node degree*: Both *MST* and *RNG* have small average node degree independently of the number n of network nodes, while the node degree in *minR* increases with n.

- *Hop diameter*: *minR* has the smallest hop-diameter and *MST* the largest, while *RNG* is in between the two.

- *Transmitting range*: *MST* has the smallest average transmitting range, while *minR* has the largest such range. The average transmitting range with *RNG* is very close to that with *MST*.

- *Biconnectivity*:[3] the *minR* graph is biconnected in all the simulated scenarios; the *RNG* topology has more than 85% of the nodes in the largest biconnected component; *MST* is a tree, so it cannot ensure biconnectivity.

---

[3]Note that connectivity is not an issue since, assuming the maximum transmit power is high enough, all the three topologies considered preserve connectivity in the worst case.

Overall, the results of the simulation-based analysis show that *RNG* is a good compromise between the goals listed above: it has relatively low logical node degree, its hop diameter is not too larger than that of the maxpower communication graph, the average node transmitting range is quite low, and a good percentage of network nodes are biconnected. Motivated by this observation, Borbash and Jennings present a protocol, called DISTRNG, for computing the *RNG* in a fully distributed and localized fashion.

Before introducing the protocol, we need the following notion of neighbor coverage:

Definition 11.2.2 (Neighbor coverage) *Consider a network node u, and one of its neighbors v. The neighbor coverage of node v, denoted by $Cov_u(v)$, is defined as the cone centered at u that spans Lune(u, v), that is, the cone of width $\overset{\frown}{aub}$, where a and b are the intersection points of the circumferences of radius $\delta(u, v)$ centered at u and v, respectively. The covered region of node u is the union set of the coverage of all its neighbors.*

The concepts of neighbor coverage and covered region are illustrated in Figure 11.6.

The DISTRNG protocol is reported in Figure 11.7. The protocol is composed of a sequence of neighbor discovery phases: initially, node *u* grows its transmit power level until a new neighbor *v* is discovered; then it adds *v* to its neighbor set, updates the covered region, and checks whether the entire 2π span is covered. If not, it increases the transmit power until a new neighbor in the not-yet-covered region is identified, and repeats the operations above. This procedure is repeated until the condition on coverage is satisfied or the maximum transmit power is reached.

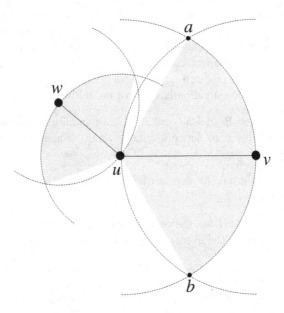

Figure 11.6 The neighbor coverage of node *v* is defined as the cone of width $\overset{\frown}{aub}$ centered at *u*. The covered region of node *u* (shaded area) is the union set of the neighbor coverages of nodes *v* and *w*.

Algorithm DISTRNG:
(algorithm for node u)

$p(u)$ is the current transmit power level of node u
P_{max} is the maximum nodes transmit power
$N(u)$ is the neighbor set of node u
$CR(u)$ is the covered region of node u
$NYCR(u)$ is the not-yet-covered region of node u

1. Initialization
    $N(u) = \emptyset$
    $CR(u) = \emptyset$
    $NYCR(u) = 2\pi$
    $p(u) = 0$

2. Computing *RNG*
    repeat until $(CR(u) = 2\pi)$ or $(p(u) = P_{max})$
        increase $p(u)$ until a new neighbor v in $NYCR(u)$ is discovered
        $N(u) = N(u) \cup \{v\}$
        $CR(u) = CR(u) \cup Cov_u(v)$
        $NYCR(u) = NYCR(u) - Cov_u(v)$

3. Finalization
    $N(u)$ is the neighbor set of node u in the final topology

Figure 11.7 The DISTRNG protocol.

In (Borbash and Jennings 2002), it is proven that the protocol reported in Figure 11.7 correctly computes the *RNG*. As in the case of CBTC, the message complexity of the protocol has not been formally analyzed nor discussed. In particular, in DISTRNG also the strategy used to increase the transmit power in the various neighbor discovery phases has a critical impact on the number of messages exchanged during DISTRNG's execution.

12

Neighbor-based Topology Control

In this chapter, we consider topology control protocols that rely on the nodes' ability to determine the number and identity of neighbors within the maximum transmitting range and to build an order on this neighbor set (based, for instance, on distance or on link quality). In a certain sense, this is the minimum amount of information needed by the nodes to build the network topology: if a node is not able to identify its neighbors, it has no clue on how to set its transmit power level. The only possible way of setting the transmit power level in this no-deterministic-knowledge scenario would be to set it at random. In this case, the amount of knowledge available to the node is represented by the probability distribution used to set the transmit power level.

In the first part of this chapter, we approach the neighbor-based topology control problem from a theoretical viewpoint, assuming that every node in the network is connected to its k closest neighbors. More particularly, we study necessary and sufficient conditions on k for generating a topology that is connected with high probability (w.h.p.). Then, we present the KNEIGH protocol introduced in (Blough et al. 2003b), which is an implementation of the above-described topology control technique based on distance estimation. Finally, we present the XTC protocol introduced in (Wattenhofer and Zollinger 2004), which is based on a more general notion of neighbor ordering, that is, link quality.

12.1 The Number of Neighbors for Connectivity

In this section, we investigate the following theoretical problem, which has been first studied in (Xue and Kumar 2004). Before introducing the problem, we need some preliminary definitions.

Definition 12.1.1 (K-neighbors graph) *Let N be a set of nodes deployed in a certain region R, with $|N| = n$. Given any k, with $0 < k \leq n - 1$, the k-neighbors graph built on N, denoted by $G_k = (N, E_k)$, is the directed graph obtained connecting each node to its k closest neighbors. Formally, the directed edge $(u, v) \in E_k$ if and only if $\delta(u, v) \leq d_k(u)$, where $d_k(u)$ is the distance between node u and its k-closest neighbor.*

Topology Control in Wireless Ad Hoc and Sensor Networks   P. Santi
© 2005 John Wiley & Sons, Ltd

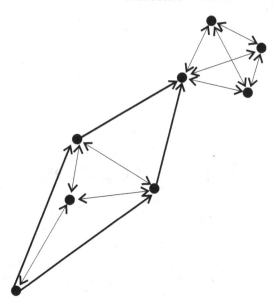

Figure 12.1 Example of node placement generating asymmetric links (bold edges) in the k-neighbors graph. Parameter k is set to 3.

Note that the neighbor relation induced by G_k is asymmetric, that is, there might exist nodes u, v such that v is a neighbor of u in G_k but u is not a neighbor of v. An example of node placement generating asymmetric links in G_k is reported in Figure 12.1.

Graph G_k can be made symmetric in two ways:

Definition 12.1.2 (Symmetric k-neighbors supergraph) *The   symmetric   k-neighbors supergraph on node set N is defined as the undirected graph $G_k^+ = (N, E_k^+)$, where the undirected edge $(u, v) \in E_k^+$ if and only if $(u, v) \in E_k$ or $(v, u) \in E_k$.*

Definition 12.1.3 (Symmetric k-neighbors subgraph) *The   symmetric   k-neighbors sub-graph on node set N is defined as the undirected graph $G_k^- = (N, E_k^-)$, where the undirected edge $(u, v) \in E_k^-$ if and only if $(u, v) \in E_k$ and $(v, u) \in E_k$.*

In other words, the symmetric supergraph of G_k is obtained by adding the reverse edge to all the unidirectional links in G_k, while the symmetric subgraph of G_k is obtained by removing all the unidirectional links in G_k. The symmetric super- and subgraph of the k-neighbors graph reported in Figure 12.1 are shown in Figure 12.2.

Definition 12.1.4 (KNEIGHCONN) *The k-neighbors connectivity problem, denoted as KNEIGHCONN, is as follows: given a set N of nodes, which is the minimum value of k such that the k-neighbors graph G_k built on N is strongly connected? We also consider two symmetric versions of KNEIGHCONN in which the connectivity requirement is either on the symmetric supergraph or on the symmetric subgraph of G_k.*

The interest in investigating the KNEIGHCONN problem is evident: by acting on a local parameter (the number k of neighbors to which every node is connected), we can 'control'

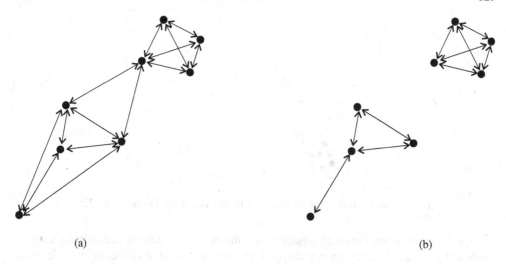

(a)   (b)

Figure 12.2 Symmetric supergraph (a) and subgraph (b) of the k-neighbors graph of Figure 12.1.

a global property, that is, network connectivity: the higher the k, the better the network connectivity. On the other hand, a low value of k is desirable for spatial reuse: the lower the k, the lesser is the number of neighbors impacted by a node's transmission, which implies increased spatial reuse. So, the optimal choice for k is to determine the minimum value such that the corresponding G_k graph is connected, that is, the KNEIGHCONN problem.

The KNEIGHCONN problem can be easily solved if some feedback mechanism can be implemented: intuitively, nodes start connecting to the closest neighbor, check for network connectivity (this is the feedback mechanism), increase the number of neighbors, and so on, until network connectivity is achieved. Unfortunately, implementing this solution in realistic scenarios is very difficult because of the fact that checking for network connectivity requires global knowledge. In general, any feedback returned to the nodes about network connectivity can be achieved only through networkwide message exchange, which, as we have thoroughly discussed in Chapter 9, should be avoided in practical scenarios.

Since having a feedback about connectivity is impractical, the only choice is to give k as an input parameter to the TC protocol, where k is chosen in such a way that connectivity is somehow guaranteed. Indeed, the connectivity requirement that we want to fulfill has a strong impact on the value of k: if we want to generate a connected graph *for every possible node placement* (worst-case connectivity), the only choice is to set $k = n - 1$. This is proven in the following theorem.

Theorem 12.1.5 *The minimum value of k that guarantees worst-case connectivity of the k-neighbors graph G_k (i.e. the minimum value of k such that the G_k graph is connected whenever the maxpower graph is connected) is $n - 1$.*

Proof. To prove the theorem, it is sufficient to show a node placement such that the maxpower graph is connected, and G_k is connected if and only if $k = n - 1$. This is the case, for instance, of the node placement reported in Figure 12.3: all the network nodes,

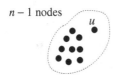

Figure 12.3 Node placement used in the proof of Theorem 12.1.5.

except for node v, are clustered together, and the nodes' maximum transmitting range is such that the maxpower graph is connected. In order to preserve connectivity, at least the node in the cluster that is closest to v, that is, node u, must have a direct link to v. Since v is the $n - 1$-closest neighbor of node u, it follows that we must set $k = n - 1$ in order to obtain a connected network.

In other words, Theorem 12.1.5 states that ensuring worst-case connectivity with closest-neighbor-based topology control is possible only by directly connecting each node to every other node in the network (i.e. the network is single hop). In real word scenarios, setting $k = n - 1$ as requested by Theorem 12.1.5 is virtually impossible: on the one hand, nodes must use relatively high transmit powers since they must be able to reach the farthest node in the network; on the other hand, network capacity is compromised: when one node in the network transmits, all the other nodes must remain silent in order not to compromise the transmission.

Given that the deterministic solution to KNEIGHCONN is not satisfactory, researchers have studied this problem in a probabilistic setting: we assume a certain node spatial distribution (e.g. the network nodes are distributed uniformly at random in a certain area), and we evaluate the value of k such that G_k is connected w.h.p.[1] In analogy with the CTR, we call the minimum such value of k as the *critical neighbor number* (CNN).

Definition 12.1.6 (Critical neighbor number) *Let N be the set of network nodes that are deployed according to some probability distribution $\mathcal{F}$ in a certain region R. The critical neighbor number (CNN) is the minimum value of k such that*

$$\lim_{n \to \infty} P(G_k \text{ is connected}) = 1,$$

where n is the number of network nodes.

A similar definition can be given with respect to the symmetric super- and subgraph of G_k.

Observe that in the study of the CNN the connectivity requirement on G_k is weakened with respect to the case of worst-case connectivity: by setting k to the CNN, we can

---

[1]We recall that a random event E_c dependent on a certain parameter c holds w.h.p., or asymptotically almost surely, if and only if $\lim_{c \to \infty} P(E_c) = 1$.

generate disconnected G_k graphs, but this occurs occasionally: the probability that G_k is disconnected converges to 0 as the size of the network increases. As we shall see, weakening the connectivity requirement on G_k has a positive effect on k, which can be considerably reduced with respect to the value of $n - 1$ needed to ensure worst-case connectivity.

Determining the CNN is a very difficult problem, which has been partially solved only recently. The difficulty with respect to the CTR problem arises from the fact that, contrary to what happens with the CTR, nodes in general have different transmitting ranges, and asymmetric links might occur. To partially mitigate the difficulty of the problem (and also motivated by the fact that unidirectional links are undesirable), Xue and Kumar considered the symmetric supergraph of G_k, and proved the following theorem:

Theorem 12.1.7 (Xue and Kumar 2004) *Assume that n nodes are placed uniformly at random in $[0, 1]^2$, and let G_k^+ be the symmetric supergraph of the k-neighbors graph built on these nodes. There exist two constants c_1, c_2, with $0 < c_1 < c_2$, such that*

$$\lim_{n \to \infty} P(G_{c_1 \log n}^+ \text{ is disconnected}) = 1,$$

and

$$\lim_{n \to \infty} P(G_{c_2 \log n}^+ \text{ is connected}) = 1.$$

The authors also provide explicit values for c_1 and c_2, which are $c_1 = 0.074$ and $c_2 = 5.1774 + \varepsilon$, where ε is an arbitrarily small positive constant.

Theorem 12.1.7 gives upper and lower bounds to the CNN in case of uniformly distributed nodes, as stated by the following corollary:

Corollary 12.1.8 *Assume that n nodes are placed uniformly at random in $[0, 1]^2$, and let G_k^+ be the symmetric supergraph of the k-neighbors graph built on these nodes. The CNN for connectivity is*

$$k_C = c \log n,$$

for some constant c, with $c_1 < c \le c_2$.

Although the difference in the values for the number of neighbors that are necessary and sufficient for connectivity is quite large, Theorem 12.1.7 is very important since it states that $\Theta(\log n)$ neighbors are necessary and sufficient for connectivity w.h.p. Comparing the values of k determined in Theorem 12.1.5 and in Corollary 12.1.8, we can see the dramatic effect of the connectivity requirement on the value of k: in order to ensure worst-case connectivity, we must set k to $n - 1$; however, if we allow occasional network disconnections, the value of k can be reduced to $\Theta(\log n)$, that is, of an exponential amount.

To some extent, Xue and Kumar's Theorem contradicts the results presented in a series of papers that considered the problem of how many neighbors are desirable in a wireless multihop network (Hou and Li 1986; Kleinrock and Silvester 1978; Takagi and Kleinrock 1984). In these papers, the wireless network is modeled as a set of nodes located on the plane according to a Poisson point process, and the problem is that of maximizing the one-hop progress of a packet in the desired direction under different transmission protocols. The optimal number of neighbors was identified as six in (Kleinrock and Silvester 1978) for a slotted ALOHA protocol where all the nodes use the same transmit power. Later on, this 'magic number' was revised to eight (Takagi and Kleinrock 1984). Hou and Lu

(Hou and Li 1986) considered the same problem with variable transmit powers, and obtained the 'magic numbers' of six and eight. Other authors have considered different network models and/or different optimization objectives, deriving other 'magic numbers': three (Hajek 1983), five and seven (Takagi and Kleinrock 1984). However, the papers cited above did not consider the issue of network connectivity. Xue and Kumar's Theorem states that *as long as connectivity is considered, no magic number exists: the minimum number of neighbors needed for connectivity grows with the number n of network nodes as $\Theta(\log n)$.*

The proof of the positive part of Theorem 12.1.7, that is, that $c_2 \log n$ neighbors are sufficient for connectivity, is based on the fact (proven in (Xue and Kumar 2004)) that every node in $G^+_{c_2 \log n}$ is connected w.h.p. to every node within distance $(1 - \varepsilon)r_n$, where $r_n = \sqrt{\frac{\eta \log n}{\pi n}}$, ε is an arbitrary constant in $(0, 1)$, and η is a constant that depends on ε. In other words, this means that the communication graph $G^{(1-\varepsilon)r_n}$ generated by the $(1 - \varepsilon)r_n$-homogeneous range assignment is a subgraph of $G^+_{c_2 \log n}$ (asymptotically, for $n \to \infty$). Hence, results concerning the CTR for connectivity can be applied, proving that $G^+_{c_2 \log n}$ is also connected w.h.p. Blough et al. observed in (Blough et al. 2003b) that, since the distance relation is obviously symmetric, the exactly same proof can be applied to show that the $G^{(1-\varepsilon)r_n}$ is a subgraph of $G^-_{c_2 \log n}$ too. In other words, $c_2 \log n$ neighbors are sufficient to have the G^-_k graph connected w.h.p. Finally, we observe that G^-_k is a subgraph of G^+_k for any k, which implies that $k > c_1 \log n$ is a necessary condition for the connectivity w.h.p. of G^-_k. Thus, we can state the following corollary, which partially characterizes the CNN for the symmetric subgraph of G_k.

Corollary 12.1.9 *Assume that n nodes are placed uniformly at random in $[0, 1]^2$, and let G^-_k be the symmetric subgraph of the k-neighbors graph built on these nodes. The CNN for connectivity is*

$$k_C = c \log n,$$

for some constant c with $c_1 < c \le c_2$.

The upper bound on the CNN of G^-_k (and, hence, the same bound for G^+_k) has been recently improved by Wan and Yi in (Wan and Yi 2004) to $\beta e \log n$, where $e \approx 2.718$ is the natural base and β is an arbitrary constant greater than 1. Indeed, the authors proved an even stronger result, showing that the same condition on the number of neighbors is sufficient for ensuring h-connectivity w.h.p., where h is an arbitrary constant with $h \ge 1$.

Another extension to Xue and Kumar's Theorem has been presented in (Blough et al. 2003a), generalizing the result to square deployment regions of arbitrary side. In fact, Theorem 12.1.7 holds under the assumption that the deployment region is fixed (it is the unit square), and the number of nodes grows to infinity. In other words, it can be applied only to *dense* ad hoc networks, where the number of nodes per unit area is quite large. Blough et al. showed that the same result holds for *sparse* networks also, and for *arbitrary* network densities in general. This generalization is important since it formally proves that *it is only the number n of nodes in the network, and not the area on which the network is deployed, that determines the CNN.*

The minimum number of neighbors needed for connectivity has been investigated in a more practical setting also. In particular, Blough et al. in (Blough et al. 2003b) and in (Blough et al. 2003a) evaluated the CNN by means of extensive simulation. In the simulation

Table 12.1 Critical neighbor number for different values of n. k_{asym} is the CNN for the G_k graph and k_{sym} is the same number for the symmetric subgraph of G_k. Here, the CNN is defined as the minimum value of k that guarantees that the generated topology is connected with probability at least 0.95

n	k_{asym}	k_{sym}	n	k_{asym}	k_{sym}
10	6	6	100	9	9
20	8	8	250	9	9
25	8	8	500	9	9
50	8	9	750	9	10
75	9	9	1000	10	10

setting, the CNN is defined as the minimum value of k that guarantees that the generated topology is connected with probability at least 0.95. Blough et al. considered both the G_k graph and its symmetric subgraph G_k^-. Note that the CNN in case of G_k^- might be higher than that of G_k because the asymmetric link removal might disconnect the graph (see Figures 12.1 and 12.2). The asymmetric CNN, k_{asym}, and the symmetric CNN, k_{sym}, for different values of n are reported in Table 12.1.

As seen from the table, setting $k = 9$ provides connectivity w.h.p. for a wide range of network sizes (from 50 to 500 nodes). Another important observation concerns the fact that k_{asym} and k_{sym} have the same value, except for a few cases in which $k_{sym} = k_{asym} + 1$. This fact confirms that, as predicted by Corollaries 12.1.8 and 12.1.9, removing all the asymmetric links has asymptotically negligible influence on network connectivity.

To gain more insights on neighbor-based connectivity, Blough et al. have also performed a set of simulations in which the connectivity requirement on G_k^- is weakened. In particular, it is requested that at least 95% of the network nodes are in the same connected component w.h.p. (here, w.h.p. means with probability at least 0.95). We call the minimum value of k satisfying the above connectivity requirement wCNN, where w stands for *weak*.

The results of this experimental evaluation, which are reported in Table 12.2, are interesting: contrary to the case of CNN, wCNN shows a converging behavior as n increases; in particular, wCNN converges to 6 as $n \to \infty$.

It is interesting to compare the different values of the minimum number of neighbors for connectivity that we have characterized in this section: depending on the connectivity requirement on G_k, this value can be equal to $n - 1$ (worst-case connectivity), or to $O(\log n)$ (connectivity w.h.p.), or to 6 (most of the nodes in the largest connected component w.h.p.). Note that in this latter case we do have a magic number of neighbors, which is 6.

Before ending this section, we give one more comment about the values of wCNN and CNN reported in Table 12.2. These values indicate quite clearly (although there is no theoretical support to this claim) that the giant component phenomenon, which we have discussed in Section 4.1 in case of the CTR for connectivity, occurs in the k-neighbors graph also. In the context of neighbor-based topology control, the giant component phenomenon

Table 12.2   WeakCNN and CNN for different
values of n. Here, wCNN is defined as the minimum value of k such that at least 95% of the
network nodes are in the same connected component with probability at least 0.95

n	wCNN	CNN	n	wCNN	CNN
10	6	6	100	7	9
20	7	8	250	7	9
25	7	8	500	6	9
50	7	9	750	6	10
75	7	9	1000	6	10

can be explained as follows. Assume every node in the network establishes a connection
to its closest neighbor, then to the second closest neighbor, and so on, until connectivity
is achieved. The experimental results reported in Table 12.2 indicate that a large connected
component in G_k is formed quite soon in this closest-neighbor-connection process: w.h.p.,
most of the network nodes are in the same connected component after only six steps (i.e.
when $k = 6$), independent of n; on the other hand, if the goal is connecting *all* the nodes
in the network, then the connection process is very likely to stop after $\Omega(\log n)$ steps.

12.2   The KNEIGH Protocol

The KNEIGH protocol introduced in (Blough et al. 2003b) is a distributed implementation
of the computation of G_k^- based on distance estimation. In other words, it is assumed that
when a node u receives a message from node v, u is able to estimate (possibly with a certain
error) the distance to node v. This can be accomplished by using one of the many distance
estimation techniques proposed in the literature. Among them, we cite the following:

- *Radio signal strength indicator*: Distance is estimated by comparing the transmitted
  power at the sender (which is piggybacked in the message) with the received power at
  the receiver of the message. This technique can be implemented without any additional
  hardware on the nodes (RSSI registers are a standard feature in many wireless network
  cards (Savvides et al. 2001)), but its accuracy is bonded to the accuracy of the radio
  channel model used to predict path loss. Since path loss is very difficult to predict
  in many environments (especially in presence of buildings, obstacles, and so on), it
  turns out that RSSI-based distance estimation provides reasonable accuracy only in a
  quite idealized setting (e.g. football field with all the nodes positioned at the ground
  level) (Savvides et al. 2001);

- *Time of arrival*: Distance is estimated by comparing the time of arrival of different
  types of signals. Typically, the radio signal is used in combination with acoustic, ultra-
  sound, or infrared signals. Because of the use of different types of signals, ToA-based
  techniques provide a much better accuracy than RSSI-based mechanisms, and can be
  implemented at a reasonable hardware cost. For example, the technique proposed in
  (Girod and Estrin 2001) uses a standard PC sound card to generate an acoustic signal,

which is received by a cheap microphone. The authors show that this technique provides good accuracy (below 3%) in realistic conditions. However, accuracy drops to only 23% when the line of sight between the nodes is obstructed by heavy obstacles.

12.2.1   Protocol description

The KNEIGH protocol, which is summarized in Figure 12.4, is very simple. Initially, every node broadcasts its ID at maximum power (as usual, we assume that all the nodes have the same maximum transmit power $P_{\max}$, and that the wireless medium is symmetric). Upon receiving broadcast messages from other nodes, every node keeps track of its neighbors, storing for each of them the estimated distance (this can be done by using one of the techniques described above). After all the initial messages have been sent, every node in the network knows its neighbor set and the distance-based ordering of the neighbors. Given this information, every node computes its k-closest neighbors list KN, and broadcasts this information at maximum power. By exchanging neighbor lists, nodes are able to determine the set of symmetric neighbors[2] and to exclude the asymmetric neighbors from KN. At the end of the protocol execution, $KN(u)$ contains the list of neighbors of node u in the final topology G_k^-, and the (broadcast) transmit power of node u is set to the minimum value needed to reach the farthest node in $KN(u)$. Note that this value can be computed given the received signal strength of the messages sent by the farthest node in $KN(u)$.

Blough et al. proved the following properties of KNEIGH (Blough et al. 2003b):

– *Correctness*: If all the nodes use the same maximum transmit power $P_{\max}$ and the wireless medium is symmetric, then algorithm KNEIGH correctly computes the G_k^- graph. To be precise, KNEIGH computes a subgraph of G_k^-, where every node is connected to its k closest neighbors, or to the maximum possible number of neighbors within the maximum transmitting range.

– *Connectivity*: Under the assumption that nodes are distributed uniformly at random in a square of arbitrary side, and that $P_{\max}$ is chosen in such a way that the maxpower communication graph is connected w.h.p., and setting k to the CNN as indicated in Corollary 12.1.9, the topology generated at the end of KNEIGH execution is connected w.h.p. For practical purposes, given the number n of network nodes, k can be set to k_{sym} as indicated in Table 12.1.

– *Bounded physical node degree*: The physical node degree of any network node at the end of KNEIGH's execution is upper bounded by k.

– *Termination*: Under the assumption that the time interval between the instants at which the first and the last node in the network broadcast the ID is at most Δ, for some $\Delta > 0$, and assuming a randomized scheduling to send messages, the protocol terminates correctly after a certain finite time T, which depends on n and Δ.

– *Message complexity*: Since every node in the network sends exactly two messages, the total number of messages exchanged by KNEIGH is $2n$.

---

[2]Two nodes are symmetric neighbors if and only if they appear in each other's KN list.

Algorithm KNEIGH:
(algorithm for node u)

k is the target number of neighbors (input parameter)
P_{max} is the maximum node transmit power
$N(u)$ is the neighbor set of node u
$KN(u)$ is the k-closest neighbor set of node u
$p(u)$ is the final (broadcast) transmit power of node u

1. Initialization
    $N(u) = \emptyset$
    $KN(u) = \emptyset$

2a. ID broadcast
    send message (u, P_{max}) at transmit power P_{max}

2b. Neighbors detection
    upon receiving message (v, P_{max}) from node v
        $N(u) = N(u) \cup \{v\}$
        estimate distance to v and store this information

3. Wait for stabilization time

4. Compute the k-closest neighbors list
    order the nodes in $N(u)$ according to the estimated distance
    $KN(u) =$ first k nodes in $N(u)$ – (all the nodes if $|N(u)| \leq k$)

5a. Neighbor list broadcast
    send message $(u, KN(u))$ at transmit power P_{max}

5b. Neighbor lists reception
    upon receiving message $(v, KN(v))$ from node v
        store this information

6. Wait for stabilization time

7. Symmetric neighbor list computation
    for each $v \in KN(u)$ do
        if $u \notin KN(v)$ then $KN(u) = KN(u) - \{v\}$

8. Transmit power computation
    $p(u) =$ minimum power level needed to reach the farthest node in $KN(u)$

Figure 12.4 The KNEIGH protocol.

Algorithm KNEIGH – Optimization Stage:
(algorithm for node u)

$KN(u)$ is the symmetric k-closest neighbor set of node u
for each $v \in KN(u)$, $P(u, v)$ is the minimum transmit power needed to reach v
$P(u, v)$, for each $v \in KN(u)$, is included in the message sent at step 5a. of KNEIGH
$p(u)$ is the final (broadcast) transmit power of node u

1. Initialization
    Sort nodes in $KN(u)$ for increasing value of $P(u, v)$
    let $v_1, \ldots, v_h$, with $h \leq k$, be the resulting ordering

2. Energy-inefficient edge removal
    for $i = 1$ to h do
        check whether $\exists v_j$, with $j < i$, such that $P(u, v_j) + P(v_j, v_i) \leq P(u, v_i)$
        (note that $P(v_j, v_i)$ has been received by u together with the list $KN(v_j)$
        sent by node v_j)
        if yes, remove v_i from $KN(u)$, and set $P(u, v_i)$ to $P(u, v_j) + P(v_j, v_i)$

3. Transmit power computation
    $p(u) =$ minimum power level needed to reach the farthest node in $KN(u)$

Figure 12.5 The optimization stage of the KNEIGH protocol.

Blough et al. also presented an optimization stage that can be applied at the end of KNEIGH. The philosophy of this optimization is the same as in CBTC, that is, to remove energy-inefficient links from the final topology. The optimization stage, which is reported in Figure 12.5, can be executed locally at each node with no further message exchange. The idea is very simple: edge (u, v_i) is removed from the final topology if and only if there exists a third node v_j that is a symmetric neighbor of both u and v_i, such that sending a message along the two-hop path $\{u, v_j, v_i\}$ is more energy efficient than the direct communication from u to v_i. Note that only edges that are part of a 'triangle' are removed, that is, this optimization does not reduce network connectivity. Furthermore, given that nodes u, v_j and v_i are symmetric neighbors, and all of them execute the same optimization protocol, node u removes node v_i from $KN(u)$ if and only if node v_i removes node u from $KN(v_i)$, that is, the optimization stage preserves symmetry.

The authors of (Blough et al. 2003b) evaluated the average-case performance of KNEIGH through simulation, comparing it with the case of no topology control (i.e. the network topology is the maxpower communication graph) and with CBTC. In case of KNEIGH and CBTC, Blough et al. compared the protocols both with and without the final optimization phase implemented. The metrics considered to evaluate the performance of the protocols were (i) the average (broadcast) transmit power level of the nodes in the final topology, which is a measure of the energy efficiency of the protocols; and (ii) the average physical node degree, which is a measure of the expected interference. The simulation results show that KNEIGH after optimization is about 20% more energy efficient than CBTC after optimization. If optimizations are not implemented, the gap in favor of KNEIGH is even larger. As for the average physical node degree, KNEIGH without optimization performs much better than CBTC without optimization, while the two protocols have essentially the same performance if optimizations are implemented.

A final simulation-based investigation reported in (Blough et al. 2003b) concerns the effect of errors in distance estimation on KNEIGH's performance. The authors considered realistic error models for both RSSI- and ToA-based techniques, and evaluated the impact of errors on the connectivity of the final topology. The simulation results show that errors in distance estimation have a very limited impact on the network topology, especially in case the relatively accurate ToA-based distance estimation technique is used. In other words, by setting k as in Table 12.1, the topology generated by KNEIGH is connected w.h.p. even if the distance estimation is relatively imprecise.

12.2.2 Discussion

The KNEIGH protocol enjoys several nice features: it is very simple, it is based on relatively 'low-quality' information (distance between nodes), it is lightweight (only $2n$ messages are exchanged in the network), and it generates a topology with a nontrivial upper bound on the physical node degree, which, as we have seen, is fundamental to maintain a relatively low level of interference in the network. However, contrary to all the topology control protocols presented so far, KNEIGH *does not preserve network connectivity in the worst case*. Unfortunately, as we have discussed in Section 9.3, there is no way of guaranteeing connectivity in the worst case and limited physical node degree at the same time. So, if the emphasis is on limiting the physical node degree as it is in KNEIGH, guaranteeing connectivity w.h.p. is, in a sense, the best one can hope for.

As compared to CBTC, KNEIGH displays about 20% better performance in terms of energy cost, and slightly better performance in terms of average physical node degree. The reason for this performance gap in favor of KNEIGH stems from the fact that some nodes in CBTC (especially nodes lying on the boundary of the deployment region) use relatively high transmit power to reach at least one neighbor 'in every direction': on one hand, this feature of CBTC ensures that the generated topology preserves worst-case connectivity; on the other hand, it produces a relatively less energy- and interference-efficient topology as compared to KNEIGH.

12.3 The XTC Protocol

The XTC protocol presented in (Wattenhofer and Zollinger 2004) is a neighbor-based topology control algorithm that, contrary to KNEIGH, preserves worst-case connectivity. This better performance in terms of connectivity with respect to KNEIGH is counterbalanced by a worse performance in terms of physical node degree, which can be as high as $n - 1$ (see Theorem 9.3.3). We recall that KNEIGH generates graphs with physical node degree upper bounded by k, where $k \in O(\log n)$.

In a certain sense, XTC can be considered as a generalization of KNEIGH: similar to KNEIGH, nodes first establish an order on their neighbor nodes; then, they exchange information about the neighbor orders; finally, they compute their local view of the final network topology. The main differences between XTC and KNEIGH are the following:

1. The neighbor order is based on the concept of 'link quality', rather than distance as it was the case in KNEIGH.

2. When exchanging neighbor lists and computing the final topology, nodes consider the entire neighbor set, and not the first k elements in the order as was the case in KNEIGH.

Let us comment more on issues (1) and (2). The use of distance in KNEIGH was motivated by the need of upper bounding the physical node degree, which is defined as the number of nodes within a node's transmitting range: by setting the transmit power to the minimum level needed to reach k neighbors, we ensure that the physical degree of a node is k. Is it necessary that these neighbors are the closest ones, in term of Euclidean distance? The answer is *no* if we consider the correctness of KNEIGH, but it is *yes* if we want to have a probabilistic guarantee on the connectivity of the generated topology: the most natural way to prove that the topology computed by KNEIGH is connected w.h.p. is by using the theory of k-neighbors graph, which is based on the concept of Euclidean distance. In case of XTC, the focus is on ensuring worst-case connectivity, which, as we know, implies that the physical node degree cannot be bounded. As a consequence of this, the number of neighbors within a node's transmitting range is not an issue, and the notion of Euclidean distance is no longer necessary since connectivity is guaranteed in the worst case. Thus, nodes can use a more general and practical notion such as link quality to order their neighbors. We remark that, under particular circumstances (e.g. flat and unobstructed environment), the order induced on neighbors by link quality might coincide with the order induced by distance. Finally, note that nodes in XTC must exchange their entire neighbor set (issue (2)) since otherwise worst-case connectivity cannot be ensured.

12.3.1 Protocol description

Before presenting the protocol, we need some notation. Let us consider a certain node u, and let $N(u)$ be its neighbor set (i.e. the set of nodes within u's transmitting range at maximum power). In the following, we denote the order relation on $N(u)$ by $\prec_u$; in particular, $w \prec_u v$ means that node w precedes node v in the ordering of node u. In terms of link quality, $w \prec_u v$ indicates that link (u, w) has relatively higher quality than link (u, v). As usual, we assume that all the nodes in the network have the same maximum transmit power $P_{\max}$ and that the wireless medium is symmetric.

As anticipated above, the XTC protocol (which is summarized in Figure 12.6) is very simple: initially, every node in the network establishes an order $\prec$ on the neighbor set. The way this can be accomplished depends on the criterion used to establish link quality: it can be defined in terms of the received signal strength, or in terms of packet delivery ratio, and so on. To simplify the presentation of XTC, in the protocol specification reported in Figure 12.6, it is assumed that link quality is determined by the strength of the received signal. With this assumption, establishing the neighbors order is very simple: each node sends a beacon at maximum power; when a neighbor receives the beacon, it measures the received signal strength (this can be easily done since wireless cards are typically equipped

Algorithm XTC:
(algorithm for node u)

P_{max} is the maximum node transmit power
$N(u)$ is the neighbor set of node u
$DN(u)$ is the set of discarded neighbors
$FN(u)$ is the neighbor set of node u in the final topology
$p(u)$ is the final (broadcast) transmit power of node u

1. Initialization
    $N(u) = \emptyset$
    $DN(u) = \emptyset$
    $FN(u) = \emptyset$

2a. ID broadcast
    send message (u, P_{max}) at maximum transmit power P_{max}

2b. Neighbors detection
    upon receiving message (v, P_{max}) from node v
        $N(u) = N(u) \cup \{v\}$
        determine the received signal strength and store this information

3. Exchange ordered neighbors list
    after all the messages from neighbors have been received
        order nodes in $N(u)$ according to the received signal strength

    3a. send message $(u, N(u))$ at maximum transmit power P_{max}

    3b. upon receiving ordered neighbors list $(v, N(v))$ from node v, store this
        information

4. Determine final neighbor set
    after the ordered neighbor lists from all the nodes in $N(u)$ have been received
        for each $v \in N(u)$, in decreasing order of link quality
            if $(\exists w \in DN(u) \cup FN(u)$ such that $w \prec_v u)$
            then $DN(u) = DN(u) \cup \{v\}$
            otherwise $FN(u) = FN(u) \cup \{v\}$

5. Transmit power computation
    $p(u) = $ minimum power level needed to reach the farthest node in $FN(u)$

Figure 12.6 The XTC protocol.

with a RSSI register) and orders its neighbor set accordingly. So, exchanging n messages overall (every node sends one message at maximum power) is sufficient to compute the neighbors order.

Once the neighbor set has been computed and ordered, every node broadcasts the ordered neighbor list at maximum power (this also requires sending n messages overall).

The final step of XTC, that is, the computation of the final topology, can be done locally at each node, with no further message exchange. To compute the network topology, node u considers its neighbors in decreasing order of link quality: when considering a certain neighbor v, it checks whether there exists a node w with $w \prec_u v$ (i.e. with better link quality) such that $w \prec_v u$. Note that this check can be done locally at node u, which knows v's neighbor order (because v is a neighbor of u). If the above condition is satisfied, edge (u, v) is discarded; otherwise it is included in the set $FN(u)$, which contains the neighbors of u in the constructed network topology. After all the nodes in $N(u)$ have been processed, set $FN(u)$ is the u's local view of the network topology produced by XTC, which we call G_{XTC}.

12.3.2   Protocol analysis

In (Wattenhofer and Zollinger 2004), Wattenhofer and Zollinger investigate the properties of the G_{XTC} graph in two different settings. First, they consider the more general setting in which $\prec$ is an arbitrary link quality–based order; then, they consider a quite idealized setting in which the link quality–based order coincides with the distance-based order (as discussed above, this might happen if, for instance, the environment is flat and unobstructed), and proves additional properties of the G_{XTC} graph.

Let us start with the more general setting in which the measure used to determine link quality is arbitrary. The first property considered by Wattenhofer and Zollinger is symmetry: in particular, they show that the neighbor relation induced by G_{XTC} is symmetric:

Theorem 12.3.1 (Wattenhofer and Zollinger 2004) *Let $G = (N, E)$ be the maxpower communication graph, and let $G_{XTC} = (N, E_{XTC})$ be the topology computed by the XTC protocol, that is, edge $(u, v) \in E_{XTC}$ if and only if $v \in FN(u)$ at the end of XTC's execution. The G_{XTC} graph is symmetric, that is, edge $(u, v) \in G_{XTC}$ if and only if $(v, u) \in G_{XTC}$. Equivalently, at the end of XTC's execution $v \in FN(u)$ if and only if $u \in FN(v)$.*

Proof. Assume for the sake of contradiction that $v \in FN(u)$, but $u \in DN(v)$. Since $u \in DN(v)$, there exists node $w \in N(v)$, with $w \prec_v u$, such that $w \prec_u v$. Let us now consider the time at which node u processes neighbor v; since $w \prec_u v$, this implies that w has already been processed by u, that is, $w \in DN(u) \cup FN(u)$ when u processes v. In turn, this implies that when the condition on node v is checked, v must be inserted in $DN(u)$; in fact, node w is such that $w \in DN(u) \cup FN(u)$, and $w \prec_v u$. This contradicts the initial assumption that $v \in FN(u)$, and the theorem is proven.

Note that, contrary to the case of CBTC in which the symmetry of the final topology is enforced by adding the reverse edge in the unidirectional links (or by removing all the unidirectional links), in XTC it is the neighbor relation itself that is symmetric, so adding or removing unidirectional links is not needed.

The following theorem proves that XTC preserves connectivity in the worst case.

Theorem 12.3.2 (Wattenhofer and Zollinger 2004) *Let* $G = (N, E)$ *be the maxpower communication graph, and let* $G_{XTC} = (N, E_{XTC})$ *be the topology computed by the XTC protocol.* G_{XTC} *is connected if and only if* G *is connected.*

Proof. Proving the 'only if' part is immediate since G_{XTC} is a subgraph of G.

To prove the reverse implication, assume for the sake of contradiction that G is connected, but G_{XTC} is not connected. This implies that there exists at least one node pair that is connected in G but is not connected in G_{XTC}. Among all such pairs, consider the pair u, v of minimum cost, where the cost of pair u, v is given by the cost (in terms of link quality) of the minimum path between u and v in G. To break possible ties, we consider the lexicographical order of the node IDs. Since u, v is the 'disconnected' pair in G_{XTC} of minimum cost, it follows that u and v are one-hop neighbors in G. Otherwise, there exist nodes w, z in the minimum path connecting u and v in G such that nodes w and z are disconnected in G_{XTC}, and the cost of w, z is strictly less than the cost of u, v – contradiction. Since $(u, v) \in G$, it follows that $v \in N(u)$, and that v is included in $DN(u)$ when it is processed by node u (in fact, edge (u, v) is not in G_{XTC}). In turn, this implies that there exists node w, with $w \in N(u)$, such that $w \prec_v u$, that is, node w is also a neighbor of node v. It is immediate to see that we must have $w \in FN(u)$ (and $w \in FN(v)$), since otherwise u, v would not be the pair of minimum cost that is connected in G and disconnected in G_{XTC}. This leads to a contradiction since $(u, w) \in E_{XTC}$ and $(w, v) \in E_{XTC}$ implies that u and v are connected in G_{XTC}, contrary to our initial assumption.

Concerning message complexity, XTC can be classified as a lightweight protocol, since its computation requires exchanging $2n$ messages (under the assumption that link quality is measured using received signal strength).

In case the link quality–based order coincides with the neighbor order based on Euclidean distance, G_{XTC} satisfies some additional property. In particular, in (Wattenhofer and Zollinger 2004), it is proved that G_{XTC} has logical node degree at most 6, it is planar, and that it is a subgraph of the RNG. More particularly, it is shown that if the node placement is such that no node has two or more neighbors at the same distance (as it is likely the case in presence of random node distribution), G_{XTC} is exactly the RNG. Thus, the algorithm reported in Figure 12.6 can be considered a distributed implementation of the computation of the RNG, as it is the DISTRNG protocol of Figure 11.7. Note that a notable feature of XTC as compared to DISTRNG is that it does not require directional information, which is typically provided using expensive directional antennas.

Summarizing, the XTC protocol

- computes a topology that contains only bidirectional links;

- preserves worst-case connectivity;

- is lightweight;

- in case the link quality–based order coincides with the distance-based order, the XTC protocol

    - produces a planar topology with logical node degree at most 6;

    - produces a subgraph of the RNG.

13

Dealing with Node Mobility

In the previous chapters, we have presented several distributed topology control (TC) protocols, based on different approaches (location-based, direction-based, and neighbor-based TC). When describing the protocols and analyzing their properties, we implicitly assumed that *the network nodes were stationary*. Indeed, *node mobility* is a prominent feature of ad hoc networks: in most application scenarios, the wireless devices that form the network, or at least a significant percentage of them, are mobile. This is the case, for instance, of ad hoc networks used to deliver traffic information (here, vehicles can be seen as network nodes), or to provide ubiquitous Internet access (here, portable devices carried by humans can be used to increase service coverage), or in the delivery of location-aware information (as in the previous example, humans carrying a wireless device can be seen as network nodes). In some cases, node mobility is present in wireless sensor networks also: for instance, if sensors are deployed on the surface of the ocean to monitor, say, the water temperature, we can expect that they are carried around by ocean flows.

So, it seems that current literature on TC has ignored one of the most important features of ad hoc and sensor networks, that is, node mobility. Is this fact true? As we shall see, the answer to this question is 'in part, yes': although some TC techniques explicitly designed for mobile networks have been introduced, many fundamental issues related to applying TC in mobile networks have not been addressed yet.

Leaving the discussion of open research issues related to the application of TC in mobile networks to Chapter 15, in this chapter we review the current state of the art on this topic. We start by revisiting the design guidelines discussed in Chapter 9 in the context of mobile networks. Then, we discuss the effect of node mobility on the value of the CNN, which, as we have seen in the previous chapter, is a fundamental network parameter in neighbor-based TC. In the last section, we present some of the TC protocols (or reconfiguration procedures of known protocols) that have been proposed in the literature to deal with node mobility.

13.1   TC Design Guidelines with Mobility

As we have discussed in Chapter 9, a TC protocol should be designed according to several guidelines, which we summarize below:

1. fully distributed and asynchronous implementation;

2. construct the topology using only local information;

3. build a topology that preserves the original network connectivity (at least w.h.p.) using only bidirectional links;

4. construct a topology with small physical node degree;

5. use relatively 'low-quality' information to build the topology.

   Let us consider these design guidelines in the context of mobile networks. A first comment is about what we mean by mobile network. It is clear that different types of node mobility may occur in ad hoc networks, ranging from highly mobile networks (e.g. vehicular ad hoc networks, where node velocity can be above 100 km/h), to networks in which node mobility is extremely low (say, if a sensor network is used to monitor the movement of turtles). Since the latter type of mobility is well approximated by stationary networks, in this chapter we are concerned with networks in which node mobility is at least moderate. In other words, we want to discuss what changes in the picture that we have drawn in Chapters 9–12 when the assumption that node positions do not change during the execution of the TC protocol and for a certain period of time after its execution is dropped.

   Guidelines (1) and (2), which were important in case of stationary networks, become *vital* in the context of mobile networks: centralized approaches and/or solutions that require the exchange of global information are impractical when node mobility is moderate to high, unless the network is extremely small (say, up to 10–15 nodes that are at most 1–2 hops away from each other). In fact, the propagation of networkwide information requires a lot of time, and the consequence of using global information to construct the communication topology is building a topology based on stale information. If the information used to compute the topology is outdated, nodes are likely to experience frequent link errors when communicating with other nodes, which in turn causes the execution of route recovery procedures and/or reexecution of the TC protocol. Thus, a considerable portion of the (scarce) network bandwidth is used by control packets, which are exchanged to continuously update the network topology and the routing information. In the most pessimistic scenario, we are in a situation in which the network topology never stabilizes, and almost the entire bandwidth is wasted for exchanging of control packets.

   In order to avoid the problems described above, the protocol used to build the network topology must be *fast*, so that it can catch up with the changes that are going on in the network. How much fast the TC protocol must be depends on the rate of node mobility: the higher the mobility rate, the faster the TC protocol has to be. To be fast, a protocol should *exchange relatively few messages with neighbor nodes* (exchanging messages with faraway nodes is time consuming), and should *execute a simple algorithm to compute the neighbor set* based on the information contained in the exchanged messages.

   Given this strong design constraint (exchange few messages with neighbor nodes, and execute a simple algorithm to compute the topology), it is clear that having ambitious optimization goals such as building a topology that preserves worst-case connectivity is almost impossible. Furthermore, we should consider that even if the TC protocol is very smart and efficient and builds a topology that is connected at time t a relatively small node movement at time $t + \varepsilon$ could disconnect the network anyway. Since the TC protocol cannot be executed too frequently in order to limit control message overhead (this point will be carefully discussed in the following, and in Chapter 15), it is clear that the optimization requirements (3) and (4) above must be weakened in the context of mobile networks, and intended more as guidelines in the design of a good TC heuristic. For instance, requirement (3) can be interpreted as follows: the designed protocol should keep the network, or at least a vast majority of the network nodes, connected for most of its operational time. The same applies for the physical node degree: the physical node degree should be kept as small as possible (as long as this does not impair network connectivity) for most of the node lifetime. As for the second requirement of issue (3) (symmetry), we observe that it is relatively easier to build a topology based only on bidirectional links since this can be accomplished by exchanging few localized messages and executing a simple algorithm (see, for instance, the protocols presented in Chapter 12).

Let us now comment on the quality of the information used to build the topology (issue (5)). In case of mobile networks, the TC protocol should use information that is, in a sense, 'resilient' to node mobility. We illustrate this point with an example. Consider the three types of information used in Chapters 10–12, that is, location information, directional information, and neighbor information. Let us consider two network nodes u, w, which are moving with certain velocities v_u, $v_w > 0$. Since the nodes are moving, their absolute positions change, that is, the location information of nodes u, w changes, and it changes continuously over time. However, if the nodes are moving in the same direction, their relative direction (which is the information used in direction-based TC) *does not* change. Consider now the case in which nodes are moving in such a way that their relative distance does not change too much; with this type of mobility, it is possible that the neighbor ordering used to build the topology in neighbor-based TC approaches does not change as well. This example clearly indicates that *using relatively inaccurate information* such as directional and neighbor-based information *is preferable in mobile networks*, as this type of information is likely to be less influenced by node mobility.

A final comment concerns per-packet versus periodical TC in mobile networks. We recall that in the per-packet approach a node u maintains, for each node v in its neighbor list, the transmit power to be used when sending packets to v, which is typically the minimum power needed to reach it. This way, a node can send each packet with the minimum possible energy consumption, and also spatial reuse is increased. Besides individual transmit power levels for each neighbor, every node in the network also sets a broadcast power level, which is used to send a message to all its one-hop neighbors simultaneously. Typically, the broadcast power is set to the minimum level needed to reach the farthest node in the neighbor list. In the periodical approach to TC, the management of the power levels is simplified: a node maintains only the neighbor list and the broadcast power level. Each packet is sent using the same power level, independent of the actual neighbor to which it is destined. By setting this common power level to the broadcast power, we are ensured that the messages are correctly received by the interested neighbor. The computation of the neighbor list and of the broadcast power is repeated periodically to account for changes in the network topology, from which the name periodical topology control derives.

While per-packet TC is in general more efficient in stationary networks (if certain technological problems can be solved – see Chapter 14) – and actually it is implicitly used in many TC protocols described so far, periodical TC is probably the only feasible choice in mobile networks. In fact, as we have already discussed, in the presence of node mobility the information about a neighbor position/direction that we have at time t might become stale at time $t + \varepsilon$, and TC must be intended as a heuristic to maintain a sufficiently good communication graph, rather than as an algorithm aimed at solving a certain topology optimization problem. Hence, the goal of determining for each neighbor the minimal power that is needed to communicate with it is probably too ambitious in this context because this parameter changes often as nodes move around. The considerably simpler power level management used in periodical TC is more appropriate for the mobile network scenario, and it is more in line with the philosophy of 'maintaining a reasonably good topology', which inspires distributed TC in presence of mobility.

The following example clarifies the argument that periodical TC is the right choice in mobile ad hoc networks. Assume node u can use four different power levels, which we denote as $p_0, \ldots, p_3$, and which translates into four different transmitting ranges (see Figure 13.1). Assume node u executes a certain TC protocol P, which terminates its execution at time t, and returns the neighbor list of node u, which is $N(u) = \{q, s, v, w, z\}$. Let us consider two scenarios: P uses per-packet TC (scenario (a)) and P uses periodical TC (scenario (b)). In scenario (a), node u sets a specific power level for each neighbor and the broadcast power level. The settings are as follows (see Figure 13.1):

$$\text{node } v \longrightarrow \text{power level } p_0$$

$$\text{node } w \longrightarrow \text{power level } p_1$$

$$\text{node } z \longrightarrow \text{power level } p_1$$

$$\text{node } q \longrightarrow \text{power level } p_2$$

$$\text{node } s \longrightarrow \text{power level } p_2$$

$$\text{broadcast power} \longrightarrow \text{power level } p_2 \ .$$

In scenario (b), node u only maintains the neighbor list $N(u)$ and the broadcast power level p_2, which is used to send the packets independent of which is the actual destination node.

After a certain time ε, during which P is not reexecuted, node positions are changed, as depicted in Figure 13.1(b). In scenario (a), node u experiences link failures when sending messages to nodes v, z, and s, and uses a nonminimal transmit power when sending packets to node w. This relatively high number of link failures is likely to cause relatively many route breakages, which, in turn, result in the execution of route recovery procedures and/or reexecution of the topology control P. In scenario (b), node u experiences link failures only when sending packets to node s, which migrated out of u's transmitting range at the broadcast power p_2. As a consequence, relatively less route breakages occur, and the control message overhead is reduced. Note that, in scenario (b), the only other change in u's local view of the network topology is that node r is now within u's transmitting range, establishing a new (possibly unidirectional) link. This new link will be discovered at the next execution of the TC protocol, but its presence in general does not cause a surge in control message overhead.

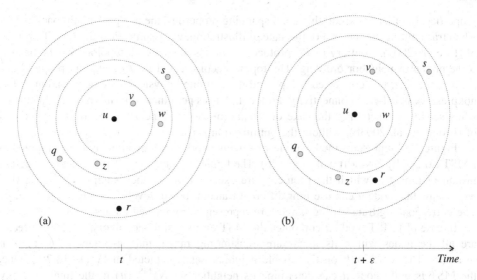

Figure 13.1 Per-packet versus periodical topology control in mobile networks. The transmitting ranges of node u at different transmit power levels are represented by the dashed circles. The topology control protocol is executed at time t, and nodes q, s, v, w and z (gray nodes) are identified as u's neighbors. At a later time $t + \varepsilon$, node positions are changed, and the power levels computed by the TC protocol are outdated.

Summarizing, a TC protocol for mobile networks should

1. Be fully distributed and asynchronous.

2. Be fast, especially if node mobility is high; in turn, this implies that the protocol should exchange few messages with neighbors, and decide its neighbor set according to a simple algorithm.

3. Generate a 'reasonably good' network topology composed of bidirectional links, that is, a topology in which most of the nodes are connected for most of the network operational time, and have a relatively small physical degree.

4. Rely on information that is relatively 'resilient' to node mobility, such as directional information or neighbor ordering.

5. Be based on the periodical approach to topology control.

13.2   TC in Mobile Networks: an Example

In this section, we present an example that summarizes the discussion on distributed TC in mobile networks of Section 13.1.

Suppose we have two TC protocols, P_1 and P_2. P_1 is based on location information, and computes a topology that preserves worst-case connectivity and has good energy spanning properties. In order to exploit this good spanning properties, the per-packet approach is used when transmitting messages. For the sake of illustration, we assume P_1 is the LMST protocol of (Li et al. 2003). Contrary to P_1, protocol P_2 builds the network topology on the basis of some notion of neighbor ordering. The topology constructed by P_2 has good properties on the average (e.g. it is connected w.h.p., and it has limited physical node degree), but it does not preserve worst-case connectivity. Protocol P_2 uses periodical transmit power adjustments when sending packets. For the sake of illustration, we assume P_2 is the KNEIGH protocol of (Blough et al. 2003b), without the optimization stage.

Figure 13.2 represents the local view of the network topology at node u as computed by LMST (a) and by KNEIGH with $k = 4$ (b). The figure depicts the node placement at a certain instant of time t at which the protocols are executed. As in the example of Figure 13.1, we assume that node u can use four different transmit power levels, denoted as $p_0, \ldots, p_3$. The corresponding transmitting ranges are represented by dashed circles.

In case of LMST, node u computes the MST on its visible neighbors, which, we recall, are all the nodes within its maximum transmitting range, that is, nodes o, q, r, s, v, w and z. The MST built on the visible neighbor set is depicted in Figure 13.2(a). Once the MST is built, node u can determine its neighbor set $N^{\text{LMST}}(u)$ in the final topology, which is composed of all the immediate neighbors of u in the local MST. In our example, $N^{\text{LMST}}(u) = \{o, v, z\}$. We recall that in LMST the constructed topology, which in general can contain unidirectional links, is made symmetric by probing each node in $N^{\text{LMST}}(u)$, and by removing the link (or adding the reverse link) in case it is unidirectional. In our example, we assume that the generated topology is made symmetric by removing unidirectional links. In order to compute its local view of G_{LMST}^- (which is the topology generated by LMST), node u must send four messages: one to broadcast its ID and position and three messages to probe the links with nodes o, v, and z. The power level settings of node u are

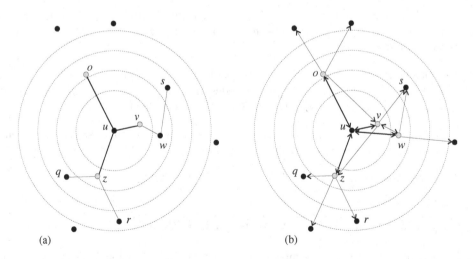

(a)   (b)

Figure 13.2 Local view at node u of the topology computed by LMST (a) and KNEIGH (b) at time t. The parameter k in KNEIGH is set to 4. The links connecting node u to its neighbors (gray nodes) are in bold.

as follows:

$$\text{node } v \longrightarrow \text{power level } p_0$$

$$\text{node } z \longrightarrow \text{power level } p_1$$

$$\text{node } o \longrightarrow \text{power level } p_2$$

$$\text{broadcast power} \longrightarrow \text{power level } p_2$$

Let us now consider the KNEIGH protocol (Figure 13.2 (b)). Node u establishes a link with its k closest neighbors. Since $k = 4$, we have that $N^{KN}(u) = \{o, v, w, z\}$. Note that all the links to nodes in $N^{KN}(u)$ are bidirectional (see the figure), so all the nodes in $N^{KN}(u)$ are retained in the final topology. In order to compute its local view of G_k^- (which is the topology generated by KNEIGH), node u must send two messages: one to broadcast its ID and a second message to broadcast its neighbor list (we recall that sending this message is necessary to identify symmetric neighbors). Since we are assuming periodical TC, node u uses the broadcast power level, which is p_2 in the example, to send packets to all the nodes in $N^{KN}(u)$.

Figure 13.3 depicts the node placement at time $t + \varepsilon$ as a result of node mobility. During this short time interval, the TC protocol is not reexecuted, which implies that node u manages the power levels as if the node placement was not changed since time t. What happens to the links used by node u?

In case of LMST, two of the three links used by u to communicate with its neighbors are broken (see Figure 13.3(a)): the link to v is broken since node u uses power level p_0 to communicate with v, and this transmit power is no longer sufficient to reach v; the link to o is broken since o migrated out of u's transmitting range at power level p_2. So, unless LMST is reexecuted, node u at time $t + \varepsilon$ can only communicate with node z, and

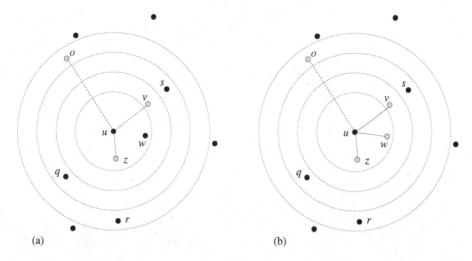

(a)   (b)

Figure 13.3 Node placement at time $t + \varepsilon$. In case of LMST (a), two of the three links of node u are broken (dashed edges). In case of KNEIGH, only one of the four links of node u is broken (dashed edge). The neighbors of u at time t are the gray nodes.

it communicates with this node using a nonminimal power level (it uses power level p_1 instead of p_0). Because of the erroneous management of the power levels at node u, many route breakages occur: in principle, most of the data flows originating, destined, or passing through node u experience packet dropping, which, in turn, causes the execution of the route recovery mechanism. As part of the route recovery mechanism, a new execution of the TC protocol might be invoked. In any case, the result is a surge of control message overhead to fix the changes in the network topology.

In case of KNEIGH, the situation is less dramatic: only one of the four links used by u (the link to node o) is broken (see Figure 13.3(b)). Thus, at time $t + \varepsilon$, node u can still communicate with nodes v, w and z. As a consequence of this, relatively less route breakages occur in the network, and the increase in control message overhead is limited.

Let us now suppose that the TC protocol is executed again at time $t + \varepsilon$. The changes in u's local view of the network topology are reported in Figure 13.4. The neighbors of u with LMST are $N^{\mathrm{LMST}}(u) = \{o, w, z\}$, and the power settings are as follows:

$$\text{node } w \longrightarrow \text{power level } p_0$$

$$\text{node } z \longrightarrow \text{power level } p_0$$

$$\text{node } o \longrightarrow \text{power level } p_3$$

$$\text{broadcast power} \longrightarrow \text{power level } p_3$$

As in the previous case, node u sends four messages to compute its neighbor set: one to broadcast its ID and position and three messages to probe the links with nodes o, w and z.

The neighbors of u with KNEIGH are $N^{\mathrm{KN}}(u) = \{q, v, w, z\}$ (note that all these nodes are symmetric neighbors of u), and the broadcast power level is set to p_2. In order to

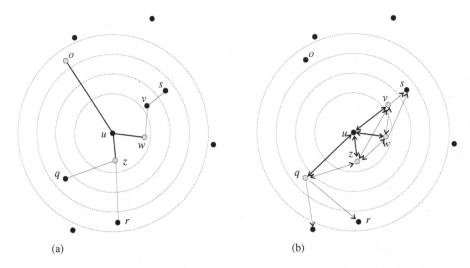

(a)   (b)

Figure 13.4 Local view at node u of the topology computed by LMST (a) and KNEIGH (b) at time $t + \varepsilon$. The parameter k in KNEIGH is set to 4. The links connecting node u to its neighbors (gray nodes) are in bold.

Table 13.1   Comparison of u's local view of the network topology at time t and $t + \varepsilon$ with the LMST and KNEIGH topology control protocols. In the entry for power levels, bc stands for broadcast power

LMST	t	$t + \varepsilon$
$N^{\text{LMST}}(u)$	$\{o, v, z\}$	$\{o, w, z\}$
Power levels	$v \longrightarrow p_0$	$w \longrightarrow p_0$
	$z \longrightarrow p_1$	$z \longrightarrow p_0$
	$o \longrightarrow p_2$	$o \longrightarrow p_3$
	$bc \longrightarrow p_2$	$bc \longrightarrow p_3$

KNEIGH	t	$t + \varepsilon$
$N^{\text{KN}}(u)$	$\{o, v, w, z\}$	$\{q, v, w, z\}$
Power level	$bc \longrightarrow p_2$	$bc \longrightarrow p_2$

compute its neighbor set, node u sends two messages: one to broadcast its ID and one to broadcast its neighbor list.

Table 13.1 summarizes u's local view of the network topology at time t and $t + \varepsilon$ with the two protocols. The neighbor set changes only slightly with both protocols: one out of three neighbors is changed with LMST and one out of four neighbors is changed with KNEIGH. The situation is considerably different if we consider the power level settings: with LMST, all the power settings are changed, including the broadcast power level; with KNEIGH, the broadcast power (which is the only power setting used by the protocol) is unchanged at level p_2.

Summing up, we can conclude the following. The topology computed by LMST is quite sparse (one of the design objectives in LMST is to reduce the node logical degree), and it is computed using location information. This implies that the generated topology, although it has good properties (e.g. it preserves worst-case connectivity), is not very resilient to node mobility: a relatively modest change in node positions is sufficient to cause a reexecution of the protocol (otherwise, many route breakages are experienced at the routing layer – see above). This problem is exacerbated by the use of per-packet transmit power adjustment. In order to prevent a surge of routing message overhead, the only possible solution is to recompute the network topology quite frequently, which also causes a certain message overhead (in the example above, node u sends four messages to compute its local view of the network topology). On the other hand, KNEIGH produces a relatively denser topology (with $k = 4$, almost all the network nodes have logical and physical degree equal to four), which has good properties in the average case (e.g. connectivity w.h.p.). The topology is built on the basis of the concept of distance ordering of the neighbors, a notion that provides more resilience to node mobility than that of local MST: relatively modest changes in the node positions are unlikely to cause a dramatic change in the neighbor order. Combined with the use of a unique power level (the broadcast power) to send packets, this implies that the network topology can be recomputed less frequently as compared to LMST, with a positive effect on the number of control messages circulating in the network. Furthermore, computing the topology with KNEIGH requires exchanging fewer messages than those needed to compute G_{LMST}^{-}: in the example above, node u sends two messages instead of four; if we consider the overall number of exchanged messages, KNEIGH exchanges $2n$ messages, while LMST exchanges $O(n^2)$ messages. This fact again plays in favor of KNEIGH, further reducing the control overhead generated by KNEIGH with respect to that generated by LMST.

Although examples of mobile networks in which the relative advantage of using KNEIGH instead of LMST is less evident can be easily built, it is virtually impossible to find an example in which using LMST in mobile networks is better than using KNEIGH. The reason for this is simple: contrary to LMST, KNEIGH combined with the use of periodical transmit power adjustments complies to most of the guidelines discussed in the previous section: KNEIGH is fast, it generates a 'reasonably good' topology composed of bidirectional links, it is based on mobility-resilient information (distance-based neighbor ordering), and it uses periodical power adjustment.

13.3 The Effect of Mobility on the CNN

In the previous sections, we have discussed in detail the guidelines that should inspire the design of TC protocols for mobile ad hoc networks, and we have argued in favor of a neighbor-based approach to TC. One protocol exploiting this type of information is the KNEIGH protocol, which we have discussed in the context of mobile networks in Section 13.2. However, we have not yet answered a fundamental question concerning the utilization of KNEIGH in mobile networks, that is: which is the appropriate setting for the desired number of neighbors k when the network is mobile? Should we use the value computed for stationary networks, or a higher (or a lower) one? Putting it more formally, which is the effect of node mobility on the CNN?

The answer to this question depends on the type of node mobility occurring in the network. In particular, the relevant parameter is the long-term node spatial distribution $\mathcal{F}_{\mathcal{M}}$ generated by a certain mobility pattern $\mathcal{M}$. Similar to the case of the CTR for connectivity (see Chapter 5), if $\mathcal{F}_{\mathcal{M}}$ is the uniform distribution (this is the case, for instance, of Brownian-like motion), we can use the value of k derived for stationary, uniformly distributed networks. If $\mathcal{F}_{\mathcal{M}}$ is not uniform, as it is the case of most of the mobility models used in the simulation of ad hoc networks, the characterization of the CNN presented in the previous chapter cannot be used.

So far, no theoretical investigation of the effect of nonuniform node mobility on the CNN has been presented in the literature. The only study in this sense is the simulation-based analysis of the CNN in presence of RWP mobility presented in (Blough et al. 2003a). The authors simulated a RWP mobile network with pause time set to 0 since this setting of the pause time corresponds to the spatial distribution that most concentrates nodes in the center of the deployment region (see (Bettstetter et al. 2003) and Section 5.1). To estimate the CNN, Blough et al. consider two connectivity requirements on the network topology: a strong connectivity requirement (the generated topology is connected with probability at least 0.95) and a weak connectivity requirement (at least 95% of the network nodes belong to the same connected component with probability of at least 0.95). We call the correspondent critical neighbor numbers as $\mathrm{CNN}_{\mathrm{RWP}}$ and $\mathrm{wCNN}_{\mathrm{RWP}}$, respectively. The simulation-based

Table 13.2   WeakCNN and CNN for different values
of the network size n, in case of stationary and RWP
mobile networks

n	wCNN	wCNN$_{RWP}$	CNN	CNN$_{RWP}$
10	6	6	6	7
20	7	7	8	9
25	7	7	8	10
50	7	8	9	12
75	7	8	9	12
100	7	7	9	12
250	7	7	9	13
500	6	6	9	13
750	6	6	10	14
1000	6	6	10	14

estimation of CNN$_{RWP}$ and wCNN$_{RWP}$ for increasing network size is reported in Table 13.2. For the sake of comparison, the table also reports the value of the CNN and of the wCNN in case of stationary, uniformly distributed network nodes.

From the table it is seen that the connectivity requirement has a considerable impact on the CNN in RWP mobile networks: in case of strong connectivity requirement, the minimum number of neighbors needed for connectivity increases significantly with n, at a higher rate than in case of stationary, uniformly distributed networks. For instance, when $n = 500$, 13 neighbors are needed to satisfy the connectivity requirement in case of mobile networks, while 9 neighbors are sufficient in stationary networks. This seems to indicate that connecting to $O(\log n)$ closest neighbors is not sufficient to generate a connected network (w.h.p.) in presence of RWP mobility. The situation is completely different if we consider the weak connectivity requirement: in this case, connecting to 6–8 neighbors is sufficient to satisfy the requirement both in stationary and in RWP mobile networks, independent of the network size. Thus, having a number of neighbors in the range 6–8 can be considered as the magical value to generate a reasonably connected communication graph in presence of RWP mobility also.

From a practical point of view, also considering the fact that we have discussed in Section 13.1, achieving full connectivity in mobile networks is a very challenging goal, and we believe that setting the value of k in the range 6–8 is the best choice.

13.4   Distributed TC in Mobile Networks: Existing Solutions

In this section, we present the few TC approaches presented in the literature that explicitly deals with node mobility.

13.4.1 The LINT protocol

The LINT protocol (Local Information No Topology) introduced in (Ramanathan and Rosales-Hain 2000) is probably the first TC protocol explicitly designed for mobile networks. The idea, which is in accordance with the guidelines of Section 13.1, is to provide a fast, simple heuristics to keep the network connected in the presence of node mobility.

LINT is a neighbor-based protocol: every node in the network tries to keep the number of its neighbors within low and high thresholds, which are centered around a certain parameter called the *desired number of neighbors*. The number of neighbors is checked at regular intervals: if it is below the low threshold, the node's transmit power is increased; if it is above the high threshold, the transmit power is decreased; otherwise, it is left unchanged. The LINT protocol is summarized in Figure 13.5.

In (Ramanathan and Rosales-Hain 2000), the authors describe a technique to adjust the transmit power, as a function of the current power level and of the actual and desired number of neighbors. An important aspect that is not considered in (Ramanathan and Rosales-Hain 2000) is how to set the value of the desired number of neighbors; however, this parameter can be configured using the characterizations of the CNN presented in Chapter 12 and in the previous section.

An important feature of LINT is the mechanism used to estimate the number of neighbors within a node's transmitting range. In LINT, a node uses locally available information provided by the routing protocol to estimate the neighbor number. In fact, routing protocols usually have a neighbor discovery mechanism, which is used to monitor the status of the links to neighbor nodes. In LINT description, it is assumed that the routing protocol returns information on bidirectional links only.

Algorithm LINT:
(algorithm for node u)

k_d is the desired number of neighbors
k_{min} and k_{max} are the low and high thresholds, centered around k_d

1. Initialization
      set the transmit power level to the initial value

2. Setting the power level
      repeat until termination
            estimate the number of neighbors, n_u
            if $n_u < k_{min}$ then
            IncrTxPower()
                  otherwise if $n_u > k_{max}$ then DecrTxPower()
            set timer TCFreq()
            wait until the timer is expired

Figure 13.5 The LINT protocol.

Estimating the number of messages using information provided by the routing protocol has the advantage of requiring no additional message overhead for TC. However, using this technique has a major drawback, that is, binding the estimation of the number of neighbors to the network traffic: if the traffic is low, or if it occurs in bursts, the routing protocol might provide little or stale information about neighbors to LINT, resulting in possibly incorrect power settings. As a consequence, this technique is indicated only for ad hoc networks in which nodes regularly exchange messages between them.

In (Ramanathan and Rosales-Hain 2000), Ramanathan and Rosales-Hain introduce another TC heuristic for mobile networks, called *LILT* (Local Information Link-state Topology). LILT is similar to LINT, with the only difference being that it is assumed that the routing protocol, besides providing information about the number of neighbors, returns to the nodes some type of global information also, such as 'the network is connected' or 'disconnected'. This type of information is available, for instance, in some link-state routing protocols, such as those presented in (Garcia-Luna-Aceves and Behrens 1995; Ramanathan and Steenstrup 1998).

LILT is composed of two procedures: the Neighbor Reduction Protocol (NRP), which is essentially LINT, and the Neighbor Addition Protocol (NAP), which is triggered whenever a link-state update indicates that the network topology has undesirable connectivity. The purpose of NAP is to override the high threshold bound on the desired number of neighbors, setting the transmit power to the maximum possible level. Indeed, the increase in transmit power levels is somehow coordinated with neighbor nodes in order to prevent an excessive reaction to the topology change (see (Ramanathan and Rosales-Hain 2000) for details).

Ramanathan and Rosales-Hain evaluate the performance of their protocols in mobile ad hoc networks through simulation, considering several performance metrics such as (i) packet delivery ratio; (ii) packet delay; and (iii) average and maximum node transmit power. They consider networks of different sizes, whose nodes move according to the random direction mobility model. The results of their experiments are interesting. Most importantly, they noticed that repeated changes in transmit power levels might increase the routing overhead because adjusting the transmit power may cause link ups/downs. If the frequency of power updates is too high, the increased routing message overhead might actually decrease the effective throughput with respect to the case of no TC, contrary to what it is expected. However, if the frequency of topology checks is correctly set, both LINT and LILT can actually increase the throughput with respect to the case of no TC. Ramanathan and Rosales-Hain also noticed that LINT *is more effective than* LILT in increasing the throughput, especially with high node densities: this is because the link-state database used by LILT to obtain information about network connectivity is often outdated, causing false alarms and unnecessary power increases. This confirms that *using global information to set up the topology in mobile networks is not only impractical, but even detrimental.*

13.4.2 The mobile version of CBTC

In (Li et al. 2001), Li et al. discuss how the CBTC protocol described in Section 11.1 can be modified to deal with node mobility. In particular, they propose a reconfiguration procedure, which is based on the Neighbor Discovery Protocol (NDP). NDP is a simple beaconing protocol used by every node in the network to tell the other nodes that it is still alive. The beacon includes the sender's ID and transmit power.

Neighbor information is updated as follows: if beacons from a certain neighbor v are not received for a certain time interval τ, then node v is considered failed (or migrated out of the transmitting range); on the other hand, a new neighbor w is detected whenever a beacon sent by w is received and no beacon was received by w for at least time τ.

The NDP protocol is used to generate events, which are dealt with by the CBTC reconfiguration protocol. Three types of events can be generated:

– *join(v)*, indicating that a new neighbor v has been detected;

– *leave(v)*, indicating that neighbor v is no longer in the node's transmitting range;

– *aChange(v)*, indicating that v's relative angle with respect to the node is changed since last beacon.

The reconfiguration protocol, which is summarized in Figure 13.6, is very simple. When a *join(v)* or *leave(v)* event is detected, the node determines whether v's removal or insertion in the neighbor set modifies the cone coverage. In case of *join(v)*, similar to the shrink back

---

Algorithm RECONFIGURECBTC:
(algorithm for node u)

NDP is the Neighbor Discovery Protocol, that generates
    join(), *leave()* and *aChange()* events

1. Initialization
    execute BASICCBTC

2. Reconfiguring the power level
    repeat until termination
        execute NDP
        case of *detectedEvent*
        join(v): insert v in the neighbor list
               compute the cone coverage
               while cone coverage is guaranteed
                 try to remove neighbors, starting from the farthest
        leave(v): remove v from neighbor list
               check condition on cone coverage
               if the condition is violated, execute BASICCBTC
        aChange(v): modify the cone coverage information
                   if the condition on coverage is violated, execute BASICCBTC
      noEvent: skip
     set timer *TCFreq()*
     wait until timer is expired

Figure 13.6 The CBTC reconfiguration protocol.

operation (see Section 11.1), the power needed to reach node v is recorded, and neighbor nodes are removed (starting from the farthest) as long as the required condition on cone coverage is satisfied. In case of *leave(v)*, it is verified whether the condition on cone coverage is impaired by v's failure; if yes, then the basic CBTC protocol is reexecuted. In case an *aChange(v)* event happens, the node updates the information about the cone coverage, and if the cone coverage requirement is not satisfied, it reexecutes BASICCBTC.

In (Li et al. 2001), Li et al. show that if the network topology ever stabilizes, then the reconfiguration algorithm eventually builds a graph that preserves the connectivity of the final network, as long as periodic beaconing is guaranteed. The authors also discuss the important topic of which transmit power to use for beaconing. They show that if the beacon is sent using the minimum power needed to reach all the nodes in the neighbor list (i.e. it is sent at the broadcast power), then the reconfiguration protocol works correctly in combination with the BASICCBTC protocol even in case the asymmetric edge removal optimization is implemented.

Part V

Toward an Implementation of Topology Control

14

Level-based Topology Control

In Chapters 10–12, we have presented several solutions to the problem of designing protocols for distributed topology control TC in ad hoc networks. The proposed solutions have
been analyzed mainly from a theoretical viewpoint: what are the properties of the generated
topology, what is the message complexity of the protocol, and so on. While this type of
analysis is important (it is, in a certain sense, an investigation of the best possible results
you can obtain with distributed topology control), the probably more important issue of
how the proposed techniques can be used in a practical setting has been almost completely
ignored in the protocols presented in Chapters 10–12.

When you think of applying distributed TC in practical settings, you have to face several
problems, mainly because of the fact that some (or many) of the assumptions on which the
protocol design is based may not hold. One such striking example is the assumption that
all the nodes have the same maximum transmit power level, which is fundamental for the
correctness of all the protocols presented in Chapters 10–12: combined with the working
hypothesis of symmetric wireless medium, this assumption ensures the correct determination
of a node's neighbor set in the maxpower communication graph. Does this assumption hold
in practice? The answer to this question depends on the application scenario. Suppose you
are in an ad hoc network used, say, for ubiquitous Internet access. In this application, the
network is composed of different types of devices (wireless access points, laptops, PDAs, cell
phones, and so on) that are used to extend the coverage area of the various wireless access
points by exploiting multihop communication. It is clear that in this scenario to assume that
all the devices have the same maximum transmit power is unrealistic: wireless access points
are likely to have a much higher maximum transmit range than, say, the maximum range
of a PDA. Let us now consider a more favorable application scenario: a sensor network
application that uses sensor nodes equipped with the same type of wireless transceiver.
Even in this situation, although the nominal maximum transmit power is the same for
all the nodes, it is likely that the actual maximum transmitting range varies considerably
between different nodes. In fact, the actual transmitting range is influenced by environmental
conditions (temperature, humidity, wind, and so on), as well as by the battery level, and so
on. In other words, *contrary to what is assumed in most of the TC approaches proposed so*

far, in a practical setting it is unlikely that all the nodes in the network will have the same maximum transmitting range.

Of course, dealing with all the issues related to the implementation of TC in a real ad hoc network is a very challenging task. Actually, most of the open research issues on TC are related to its implementation in real world scenarios (see Chapter 15).

In this chapter, we focus our attention on one such practical issue: how can the transmit power level be set in currently available wireless network cards? First, do network cards in general allow the choice of the transmit power level? If so, can we set the power level to an arbitrary value (provided it does not exceed the maximum power), or are we allowed to use only a limited number of possible transmit power levels? As we shall see, similar to the example of the equal maximum transmitting range assumption reported above, in this case also the difference between what it is typically assumed in the current TC literature and the real world is considerable.

14.1   Level-based TC: Motivations

Most of the TC solutions that approach the TC problem from a theoretical viewpoint implicitly assume that the transmit power level of a node can be set to an arbitrary level, provided it does not exceed the maximum possible power level. This is the case, for instance, of the approaches aimed at reducing the power spanning ratio of the generated network topology described in Chapter 8.

Is this assumption realistic? The answer to this question is *no*, at least with the currently available wireless cards. Indeed, most of the IEEE 802.11 wireless cards on the market do not even allow to change the transmit power level: only the maximum transmit power level can be used to communicate. Clearly, with this type of hardware, most of the TC theory is useless. Fortunately, some types of commercially available IEEE 802.11 cards, such as the CISCO Aironet cards (Cisco 2004), do allow to change the transmit power level. This is also the case of most of the wireless transceivers used in 'smart sensor' nodes, such as the Rockwell WINS (RockwellScienceCenter 2004) and the CrossBow MOTES (CrossBowTechnologies 2004). However, even if setting the transmit power level is sometimes possible, this can only be set to a limited number (typically, below 10) of predefined power levels. For instance, the CISCO Aironet 350 card can use six different power levels, corresponding to a nominal transmit power of 1, 5, 20, 30, 50, and 100 mW, respectively.

Motivated by this observation, a set of recently proposed protocols approach the TC problem by explicitly taking into account this feature of current wireless transceivers, that is, the availability of only few different transmit power levels. In the rest of this chapter, we present these solutions, which we call *level-based topology control protocols*.

14.2   The COMPOW Protocol

The COMPOW (COMmon POWer) protocol (Narayanaswamy et al. 2002) is the first proposal that appeared in the literature that explicitly deals with different node transmit power levels. Narayanaswamy et al. consider a relatively simple setting in which all the network nodes are forced to use the same transmit power level. In other words, they are considering an instance of homogeneous TC (see Section 3.3), which, as we have thoroughly discussed in Part II of this book, is the simplest possible type of TC.

The use of a common power level for all the nodes in the network is motivated by a series of practical considerations. First, using a common power level is the easiest possible way of ensuring that the generated topology is composed of bidirectional links only,[1] a feature that is very important to ease the integration of TC with upper and lower layer protocols (see Section 7.4). In particular, problems at the MAC layer arising from the use of asymmetric transmit power levels (see Section 3.4.2 and Section 15.5 for a detailed discussion of this topic) can be avoided. Finally, from a theoretical point of view, it is relatively easy to characterize the optimal common power level to be used by the network nodes.

14.2.1   The optimal common power level

Narayanaswamy et al. (2002) provide quantitative arguments to show that setting the common power to the minimum level that achieves full network connectivity is the optimal choice for increasing network capacity, reducing energy consumption, and minimizing contention at the MAC layer.

Network capacity.   Let us consider network capacity. A first interesting observation concerns the amount of potential traffic carrying capacity that is sacrificed by imposing the constraint that all the nodes must use the same transmit power level. In (Gupta and Kumar 2000) it is shown that by modeling node interference by the Protocol Model (see Section 3.1.2), the per node throughput available for a randomly chosen destination is at most $O\left(1/\sqrt{n}\right)$, where n is the number of network nodes. This upper bound on network capacity holds even if nodes are allowed to use different transmit power levels. On the other hand, a per node throughput of $O\left(1/\sqrt{n \log n}\right)$ can be achieved even in a network with randomly located nodes that use a common transmit power level. It follows that *imposing the use of a common power level does not have a dramatic impact on the potential network capacity*: it is reduced by at most a $O\left(1/\sqrt{\log n}\right)$ factor, which is asymptotically negligible as n grows to infinity.

Let us now argue in favor of using the minimum power level necessary for network connectivity. Suppose n nodes are distributed into a certain square deployment region R of area A square meters. Assume that each node in the network can transmit at W bps, and the transmitting range of each node is r meters. We model interference by the Protocol Model, which we briefly recall. In this model, a certain node v successfully receives a message from node u at distance $\delta(u, v)$ if and only if there is no other simultaneous transmitter within distance $(1 + \Delta)\delta(u, v)$ from v (see Figure 14.1). Parameter $\Delta > 0$ models the amount of interference that node v can tolerate: the smaller the Δ, the more the interference that can be tolerated. Since $\delta(u, v) \leq r$ (otherwise, v would be out of u's transmitting range), we can use a slightly stronger assumption, that is, that every node within distance $(1 + \Delta)r$ from v remains silent.

Suppose each node in the network wants to transmit data to a randomly chosen destination at a certain rate of λ bps. The question is, which setting of the transmitting range r maximizes the per node throughput λ?

---

[1]This is true under the common assumption of symmetric wireless medium.

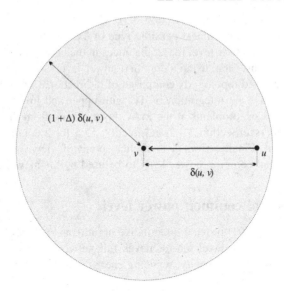

Figure 14.1 The Protocol Model for interference: in order for node v to correctly receive the packet sent by u, all the nodes within distance $(1 + \Delta)\delta(u, v)$ from v (shaded area) must remain silent.

We answer this question through a quantitative analysis. Assume that the average distance between source/destination nodes is L. Since each node has a transmitting range of r, the number of hops traversed by each packet flowing from the source to the destination is at least L/r (on the average). If we consider all the n nodes in the network, the overall number of traversed hops is at least nL/r. Since we want to obtain a per flow data rate of λ bps, at least $nL\lambda/r$ bps needs to be transmitted by all the source nodes (on the average). However, achieving this per node throughput in the network might not be possible because of interference.

Consider two simultaneous transmissions between nodes u, v and w, z; what should be the minimum distance between nodes so that the two communications do not interfere with each other? Assume u sends packets to v and w sends packets to z (see Figure 14.2). Since v must be within u's transmitting range, we have $\delta(u, v) \leq r$. On the other hand, u must be at least at distance $(1 + \Delta)r$ from z in order not to corrupt the transmission of node w. By the triangular inequality, we have that

$$\delta(v, z) + \delta(u, v) \geq \delta(u, z).$$

Combining this with the above inequalities, we obtain

$$\delta(v, z) \geq \delta(u, z) - \delta(u, v) \geq (1 + \Delta)r - r = \Delta r.$$

In other words, two transmissions can occur simultaneously only if the distance between the receivers is at least Δr. Putting it another way, we can say that, assuming r is the transmitting range, a transmission has a 'wireless footprint' $wf(r)$ of area

$$wf(r) = \pi \left(\frac{\Delta r}{2}\right)^2 = \frac{\pi \Delta^2 r^2}{4}.$$

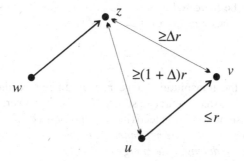

Figure 14.2 Concurrent transmissions can occur only if the distance between the receivers (nodes v and z) is at least Δr.

Since the total available area is A m^2, and each communication consumes at least area $wf(r)/4$ (this corresponds to the situation in which the node is located on the corner of R), it follows that at most

$$\frac{4A}{wf(r)}$$

transmissions can occur simultaneously. Since each node can transmit at most W bps, the total amount of bits that can be transmitted in the network per second is at most

$$\frac{4AW}{wf(r)} = \frac{16AW}{\pi \Delta^2 r^2}.$$

What is the maximum possible per node data rate that can be achieved with this network traffic carrying capacity constraint? By simple algebra, we have

$$\frac{nL\lambda}{r} \leq \frac{16AW}{\pi \Delta^2 r^2}$$

from which

$$\lambda \leq \frac{16AW}{\pi \Delta^2 nL} \cdot \frac{1}{r}.$$

So, the other parameters being fixed, the only possible way of increasing the upper bound on the per node throughput is by decreasing the transmitting range r. However, r cannot be decreased too much since otherwise network connectivity is compromised: in fact, if the network is partitioned, many nodes experience a 0 bps per node throughput (this happens whenever the source and the destination belong to different components), and the average per node throughput drops considerably. Thus, we can conclude that *setting the common power level to the minimum value that achieves full network connectivity is the optimal choice for increasing network capacity.*

Energy consumption. Let us now consider node energy consumption. We recall that the transmit power needed to send a message at distance r is proportional to r^α, where α is the path loss exponent (which is typically ≥ 2). Suppose node u has to send a packet to node v that is at distance L. Since every node has transmitting range equal to r, it follows that at least L/r hops must be traversed to send the packet. This implies that the total transmit power needed to send a message from u to v is at least

$$\frac{L}{r} \cdot r^\alpha = Lr^{\alpha-1}.$$

Thus, by reducing the transmitting range r, we can reduce the total transmit power needed to send packets between source/destination pairs. However, r cannot be decreased too much since otherwise network connectivity is compromised. Thus, we can conclude that *setting the common power level to the minimum value that achieves full network connectivity is the optimal choice for reducing node energy consumption.*

Contention at the MAC layer. Let us finally consider the expected contention at the MAC layer. We can measure the expected contention at the MAC layer by the expected number of neighbors within a node's transmitting range:[2] the higher this number, the more the neighbors that contend with the node to access the wireless channel, and the higher the expected MAC layer contention. Assuming a random, uniform node distribution, the expected number of neighbors of a node is

$$\frac{\pi r^2 (n-1)}{A}.$$

So, reducing r has the positive effect of reducing the expected number of neighbors contending for accessing the channel. However, with a lower r, the average hop length of the routes is larger, increasing the relaying burden on the nodes. Thus, the question of what is the overall effect of reducing r on the expected MAC layer contention remains open.

To answer this question, consider arbitrary source/destination nodes u, v that are L meters far away from each other. Since the transmitting range is r, at least L/r hops must be traversed to deliver the packets. What is the overall average MAC layer contention experienced to send a packet from u to v? There are L/r transmissions, each of which impact $\pi r^2 n/A$ nodes on the average. Thus, the expected overall MAC layer contention is

$$\frac{L}{r} \cdot \frac{\pi r^2 n}{A} = \frac{\pi Lrn}{A},$$

which is decreased by reducing r. Hence, we can conclude that *setting the common power level to the minimum value that achieves full network connectivity is the optimal choice for reducing contention at the MAC layer.*

14.2.2 Protocol description

In the previous subsection, we have argued in favor of setting the common power level to the minimum value that achieves full network connectivity. The COMPOW protocol is a distributed implementation of the computation and maintenance of this optimum power level.

Narayanaswamy et al. integrate COMPOW into the routing protocol, motivating this choice with the fact that connectivity is a property that can be checked only at the network

---

[2]Note that the number of neighbors within a node's transmitting range corresponds to our definition of physical node degree.

layer. COMPOW can be integrated with any routing protocol that proactively maintains routing tables at the nodes, such as the DSDV routing protocol (Perkins and Bhagwat 1994).

COMPOW is based on a very simple idea. Each node proactively maintains multiple routing tables, one for each of the power levels available on the wireless card. Routing table RT_i, corresponding to the ith power level, is built and maintained by exchanging hello messages at power level P_i. Thus, the number of entries in RT_i of node u corresponds to the number of nodes reachable from u using power level P_i. Clearly, the number of entries in RT_{max} (the routing table that corresponds to the maximum power level) gives the total number of network nodes. The optimal power level is then defined as the minimum level i such that the number of entries in RT_i equals the number of entries in RT_{max}. Once the optimal power level i is chosen, table RT_i is set as the master routing table, which is used to route packets between nodes. The COMPOW protocol is summarized in Figure 14.3. Narayanaswamy et al. have implemented COMPOW in laptops equipped with CISCO Aironet 350 cards running Linux (see (Narayanaswamy et al. 2002) for details).

14.2.3 Discussion

The COMPOW protocol has the great merit of being the first proposal that approaches the TC problem from a practical viewpoint. However, it has a major drawback, which we briefly discuss.

Algorithm COMPOW:
(algorithm for node u)

P_i is the i-th transmit power level
RT_i is the routing table corresponding to the i-th power level
RT_{max} is the routing table at maximum transmit power
$NN(u)$ is a variable containing the number of nodes in the network

1. Initialization
     start a routing daemon for each power level
     (the i-th routing daemon builds and maintains routing table RT_i)

2. Setting the power level
     repeat until termination
          $NN(u)$ = number of entries in RT_{max}
          find the minimum i such that number of entries in $RT_i = NN(u)$
          set the transmit power level to P_i
          set RT_i as the master routing table
          set timer $TCFreq()$
          wait until timer is expired

Figure 14.3 The COMPOW protocol.

COMPOW is based on the idea of using a common power level for the nodes, which is the minimum power such that the resulting network is fully connected. Although this idea is neat from a theoretical point of view, and it leads to the design of a relatively simple protocol for joint TC and routing, it has the disadvantage of setting the nodes' transmit power on the basis of a global property of the communication graph, that is, connectivity. As thoroughly discussed in Chapter 9, relying on global network properties should be avoided in distributed TC in order to reduce the message overhead needed to build and maintain the communication graph and to design a protocol that can quickly react to changes in the network topology.

The case of COMPOW well illustrates this point. As for the message overhead caused by COMPOW, we observe that proactively maintaining one routing table for each power level requires a considerable message exchange: the overhead increase with respect to the case of no TC (only one routing table is maintained) is h-fold, where h is the number of power levels used by the wireless card equipping the nodes (in practical situations, h can be as high as 10). Although the authors of (Narayanaswamy et al. 2002) argue that the message overhead due to the need of maintaining multiple routing tables is a marginal fraction of the available IEEE 802.11b bandwidth, we believe that, in practice, dealing with this considerable message overhead might be problematic.

The problem of slow reaction to topology changes in COMPOW is more dramatic. Consider the situation depicted in Figure 14.4: at a certain instant of time t, the nodes in the network use the common power level P_2, which is the minimum level necessary to

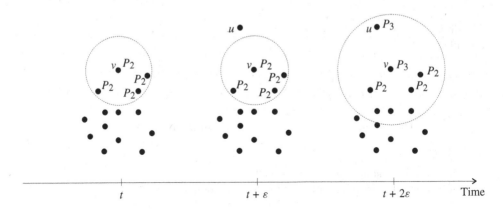

Figure 14.4 Problems caused by the slow propagation of new neighbors information in COMPOW: at time t, network nodes use the common power level P_2 (only the power level of node v and of its immediate neighbors—nodes within dashed circle—are indicated). At a later time $t + \varepsilon$, a new node (node u) joins the network, and starts exchanging *hello* messages with surrounding nodes at different power levels; node u and the closest network node, node v, can communicate only using at least power level P_3. At time $t + 2\varepsilon$, nodes u and v have updated their power level to P_3 and the master routing to RT_3, but this information has not been propagated to the other nodes yet. Consequently, node v might experience problems when communicating with its old neighbors, which still use power level P_2.

keep them connected. At a later time $t_1 = t + \varepsilon$, a new node u joins the network, and starts sending *hello* messages at the various power levels to be included in the other nodes' routing tables. In the meanwhile, u builds its own routing tables by hearing the *hello* messages sent by the other nodes. Since u is quite far from the closest node in the network (node v), it must use at least power level P_3 to be connected to the rest of the network. Thus, power level P_3 is the new common power level to be used by *all* the network nodes. Unfortunately, the propagation of this information is slow (it is propagated through routing table updates, which is a relatively slow process), and the consequence is that for a certain interval of time, which can be quite long if the network is composed of many nodes, *nodes do not use the same power level to communicate*. For instance, at a certain time instant $t_2 = t_1 + \varepsilon$, node v is aware of the new neighbor u, and, consequently, installs RT_3 as the master routing table and sets the power level to P_3. This is also the case of node u, which receives the *hello* messages sent by v only for power level P_3 or higher, and consequently sets its own power level to P_3, and uses RT_3 as the master routing table. However, all the other nodes in the network are still unaware of node u, and they continue to use the same power level and master routing table. As a consequence of this, node v might experience considerable problems when communicating with its old neighbors because of the use of asymmetric power levels.

Narayanaswamy et al. outline another potential problem of using COMPOW for joint TC and routing: the constraint of using a common power level may force most of the network nodes to use unnecessarily high transmit power. For instance, in the node configuration reported in Figure 14.4, all the network nodes will eventually use power level P_3 to communicate, but all of the nodes except u can communicate using the lower power level P_2. So, with COMPOW, a single, faraway node can cause a generalized power level increase. Interestingly, the giant component phenomenon described in Section 4.1 indicates that the unfortunate situation described above actually is very likely to occur (under the assumption of uniformly distributed nodes): we recall that the simulation results reported in (Santi and Blough 2003) show that by halving the common transmitting range with respect to the value necessary for full network connectivity we obtain a network topology in which 90% of the nodes are still connected.

Summing up, we can conclude that COMPOW can be successfully used for joint TC and routing in stationary ad hoc networks composed of relatively few nonclustered nodes. In all the other scenarios, the use of a common power level in combination with proactive routing is likely to render COMPOW impractical.

14.3 The CLUSTERPOW Protocol

The CLUSTERPOW protocol has been presented in (Kawadia and Kumar 2003) to circumvent the problem with nonhomogeneous node distribution occurring with COMPOW. We recall that with COMPOW, since all the nodes are forced to use the same power level, a single, faraway node can cause a generalized power level increase (see Figure 14.4). To solve this problem, Kawadia and Kumar release the assumption of using a common power level, and define a protocol for joint TC and routing that induces an implicit power-level-based clustering on the nodes.

14.3.1 Protocol description and properties

The CLUSTERPOW protocol displays many similarities with the simpler COMPOW protocol. As in COMPOW, every node in the network maintains separate routing tables, one for each power level. Routing table RT_i, referring to power level P_i, is maintained by exchanging *hello* messages at power level P_i. When node u has to send a message to node v, it calculates the minimum power level needed to reach node v: it is the minimum level P_i such that RT_i contains an entry for node v. Then, the packet is sent using this minimum power level. This process of calculating the minimum power level needed to reach the destination is repeated at each intermediate node in the route from the source to the destination.

The basic CLUSTERPOW mechanism described above is depicted in Figure 14.5. There are three power levels, corresponding to transmit powers of 1, 10 and 100 mW. Node u wants to send a message to node v that can be reached from u only using the maximum transmit power. As the packet gets closer to the destination (at node w_2), a lower transmit power level can be used to forward the packet. In the last hop of the path (at node w_3), the minimum power level of 1 mW can be used to forward the packet to the destination.

The example reported in Figure 14.5 outlines the following:

– CLUSTERPOW induces a hierarchical clustering on the network nodes: nodes that can reach other at power level P_i, but not at power level P_{i-1}, are in the same i level cluster. Thus, there are at most h cluster levels in the network, where h is the number of power levels. Note that CLUSTERPOW provides implicit node clustering, since there are no clusterheads nor gateway nodes. The cluster hierarchy induced by CLUSTERPOW is used for the purpose of routing only.

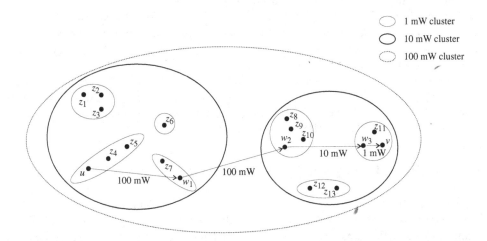

Figure 14.5 The CLUSTERPOW topology control/routing mechanism: when u wants to send a packet to node v, it uses the minimum power level P_i such that v is included in routing table RT_i. In this example, u must use power level 100 mW to send the packet, since u and v are in the same 100 mW cluster, but in different 10 mW clusters. This strategy for setting the transmit power is repeated at the intermediate nodes on the route from u to v.

- The packets flowing along the routes discovered by CLUSTERPOW are sent using *nonincreasing* transmit power levels. This is a consequence of the CLUSTERPOW forwarding policy: packets are sent through the cluster hierarchy in a top-down fashion: first, the packet is sent to other nodes in the same ith level cluster, where P_i is the minimum power level such that the sender can reach the destination. While the packet travels in the ith cluster level, it is sent using power level P_i. Once the packet reaches a node that is in the same $(i-1)$th level cluster of the destination, the transmit power level is scaled down to P_{i-1}. This process is repeated until the packet is delivered to the destination.

- Differing from COMPOW, nodes do not set one of the RT_is as the master routing table. Instead, the table used to route packets is composed of entries coming from various RT_is. In particular, the entry relative to node v in the master routing table of node u is taken from table $2RT_i$, where P_i is the minimum power level such that u can reach v.

The routing tables of node u, given the node configuration of Figure 14.5, are reported in Figure 14.6. Each entry in routing table RT_i is composed of three fields: the destination node ID, the ID of the node that is the next hop in the path to the destination, and the metric (hop count) of the path to the destination. In routing table RT_i, there is one entry for each (destination) node that node u can reach using transmit power at most P_i. In Figure 14.6, in routing table RT_i, we have reported only the entries relative to the nodes that are reachable with power level P_i but are not reachable with power level P_{i-1}. The

Routing tables of node u

1 mW routing table

Dest node	Next hop	Metric
z_4	z_4	1
z_5	z_5	1

10 mW routing table

Dest node	Next hop	Metric
z_1	z_1	1
z_2	z_1	2
z_3	z_3	1
z_6	z_5	2
z_7	z_7	1
w_1	z_7	2

100 mW routing table

Dest node	Next hop	Metric
z_8	w_1	2
z_9	w_1	2
z_{10}	w_1	2
z_{11}	w_1	3
z_{12}	w_1	2
z_{13}	w_1	3
w_3	w_1	3
v	w_1	4

Master routing table (sample entries)

Dest node	Next hop	Metric	Power lev
z_4	z_4	1	1 mW
z_2	z_1	2	10 mW
v	w_1	4	100 mW

Figure 14.6 Routing tables of node u at different power levels, and master routing table, given the node configuration reported in Figure 14.5.

master routing table is formed by joining entries from different RT_i routing tables, and by adding one additional field to each entry, namely, the power level to be used when sending packets to the destination node. For instance, the master routing table entry for destination node v is obtained by copying v's entry from routing table RT_{100} (because 100 mW is the minimum power level that allows communication between nodes u and v), and by adding the additional field 'Power level' with value 100 mW.

The CLUSTERPOW protocol for joint TC, clustering, and routing is summarized in Figure 14.7.

Note that COMPOW can be seen as a special case of the CLUSTERPOW protocol, corresponding to the situation in which network nodes are homogeneously dispersed in the environment.

Kawadia and Kumar (2003) describe how the CLUSTERPOW protocol can be used in combination with reactive routing protocols, such as AODV (Perkins et al. 2002) and DSR (Johnson and Maltz 1996). In fact, it is known that reactive routing protocols tend to perform better than proactive protocols in ad hoc networks, especially in presence of node mobility. We recall that in a reactive routing protocol source–destination paths are established on demand by a controlled flooding of *route request* messages; as route requests are flooded, the IDs of the traversed nodes are added to the request header; this way, when one of the route request messages reach the intended destination, a *route reply* message can

---

Algorithm CLUSTERPOW:
(algorithm for node u)

P_i is the i-th transmit power level
RT_i is the routing table corresponding to the i-th power level
RT_{max} is the routing table at maximum transmit power

1. Initialization
    start a routing daemon for each power level
    (the i-th routing daemon builds and maintains routing table RT_i)

2. Building the master routing table
    repeat until termination
        for each node v in RT_{max}
            find the minimum i such that v is in RT_i
            set the master routing table entry for v:
                copy the *DestNode*, *NextHop* and *Metric* field from RT_i
                set the *PowerLev* field to P_i
        set timer *TCFreq()*
        wait until timer is expired

Figure 14.7 The CLUSTERPOW protocol.

be sent back to the source node by using the reverse path. The route reply message contains the source–destination path, which will be used by the source to send the packets.

CLUSTERPOW can be used in combination with reactive routing as follows. Route request messages are sent out and forwarded at all the power levels available. When a route request message at a certain power level P_i reaches the destination, a route reply message is sent back to the source node using power level P_i (unless a route reply has already been sent using a lower power level). Once the route discovery phase is over, the source node sends the packets to the destination using the minimum power level that resulted in a successful route discovery.

Finally, we observe that CLUSTERPOW in principle might send packets into infinite loops: in fact, the master routing table is composed of entries taken from routing tables at different power levels, and, in principle, it is possible that the interaction between the routing protocols at the different power levels leads to packets getting into infinite loops. However, Kawadia and Kumar prove that, given the property that packets are forwarded to the destination with nonincreasing transmit power levels, this is not possible and CLUSTERPOW is actually loop free.

14.3.2   Implementing CLUSTERPOW

Kawadia and Kumar (2003) report their experience in implementing CLUSTERPOW on off-the-shelf components, that is, laptops equipped with CISCO Aironet 350 wireless cards.

Similar to the case of COMPOW, they implemented the protocol in the Linux environment.[3] However, Kawadia and Kumar experienced several problems in CLUSTERPOW's implementation, which de facto impeded a real testing. The major problem is that CLUSTERPOW is designed under the assumption that the wireless card is capable of changing the transmit power level on a per-packet basis (per-packet TC). Unfortunately, the CISCO Aironet 350 card only partially fulfills this requirement: there is a large latency when changing the transmit power level (Kawadia and Kumar have measured a latency in the order of 100 ms). Even worse, Kawadia and Kumar observed that frequent changes of the transmit power levels are very likely to crash the wireless card, rendering impossible CLUSTERPOW's experimentation with a significant amount of traffic.

This problem with the CISCO cards seems to be due to the fact that the firmware is written in such a way that changing the transmit power level implies resetting the card. There is no apparent technological reason for doing this. In principle, per-packet power level change should be feasible with current technology: for instance, the power is adjusted 800 times per second in cellular CDMA-based networks. So, approaches based on per-packet power level changes should be feasible in the near future, when the technical problems with the CISCO cards are fixed, and when more cards capable of dynamic transmit power change are available on the market.

Despite the problems with CLUSTERPOW's experimentation described above, Kawadia and Kumar were able to verify the correctness of their implementation. In one of the tests, they colocated five laptops running CLUSTERPOW on the same desktop, setting 100 mW as the initial transmit power level. After running CLUSTERPOW for an adequate time, the entries in the master routing tables of all the nodes had 1 mW in the Power Level field, as it was expected. At a later time, one of the five nodes was moved away from the others, so

---

[3]The source code is available online at http://www.uiuc.edu/~kawadia/txpower.html.

that it could be reached from the clustered nodes only by using transmit power 100 mW. As a reaction to this movement, the routing tables of the nodes were changed: the Power Level field of the entry relative to the outlying node in the clustered node routing tables was set to 100 mW, while the entries relative to the other nodes were left unchanged. The routing table of the outlying node had 100 mW in the Power level field, for all the entries. Thus, clustered nodes correctly used power level 1 mW for intracluster communication and power level 100 mW to communicate with the outlying node.

14.3.3 The tunneled version of CLUSTERPOW

The example of CLUSTERPOW's execution reported in Figure 14.5 shows that the protocol leaves room for further optimizations. In particular, the first 100 mW hop in the route between u and v might be replaced by a two-hop path, which consumes considerably less energy and increases spatial reuse (see Figure 14.8).

The first method to solve this CLUSTERPOW's inefficiency is to use recursive table lookup at each intermediate node in the route to the destination. For instance, in Figure 14.8, node w_1 (the next hop in u's route to node v) is recursively looked up in u's routing table, to find that w_1 is reachable from u using power level 10 mW through node z_4; in turn, z_4 is directly reachable from u using power level 1 mW. So, ultimately, the packet is sent from node u to z_4 using power level 1 mW. This recursive node lookup process is repeated at each intermediate node on the route to the destination node.

Unfortunately, the recursive lookup scheme sketched above suffers from one major problem, that is, it might lead to packets getting into infinite loops. This unfortunate situation, described in (Kawadia and Kumar 2003), is depicted in Figure 14.9. Node u wants to send a packet to node v that can be reached only by using transmit power 10 mW. The next hop in the route to v is node w_1, which is recursively looked up in u's routing tables. Node w_1 can be reached from u using transmit power 1 mW, and the next hop in the path to

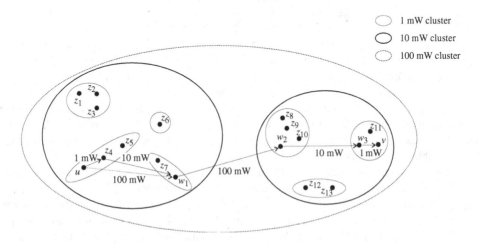

Figure 14.8 Example of CLUSTERPOW's inefficiency: the high-power hop between nodes u and w_1 might be replaced by a low-power, low-interference two-hop path.

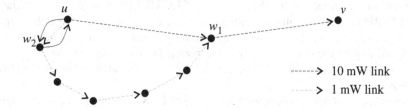

Figure 14.9 Implementing recursive lookup might lead to packets getting into infinite loops.

w_1 is node w_2. Hence, node u sends the packet to w_2, indicating that the final packet's destination is node v. The recursive lookup scheme is repeated at node w_2, which finds that the destination can be reached using transmit power 10 mW, and the next hop is node u; in turn, node u can be reached using transmit power 1 mW. So, the packet is sent back to node u, and gets into an infinite loop.

Kawadia and Kumar (2003) introduce a technique to implement a recursive lookup scheme that is free of infinite loops. The idea is tunneling the packet to its next hop using lower power levels, instead of sending the packet directly. This can be done, for instance, by using IP in IP encapsulation: while doing a recursive lookup for the next hop, the packet is recursively encapsulated with the IP address of the node that is currently looked up. When the packet reaches the next hop in the original path to the destination, it is recursively decapsulated. This version of CLUSTERPOW is called TUNNELEDCLUSTER-POW.

Figure 14.10 shows an example of TUNNELEDCLUSTERPOW's execution in the same node configuration as in Figure 14.9. The next hop in the original path from node u to node v is w_1; in turn, w_1 can be reached from u through a 1 mW path, and the next hop in this path is node w_2. Before sending the packet to w_2, w_1's ID is encapsulated in the original message. When the packet arrives at w_2, it is sent to the next hop in the path from node w_2 to node w_1, that is, node w_3. When the packet arrives at w_1, which is the next hop in the original path from u to v, the packet is decapsulated, removing w_1's ID from the packet header.

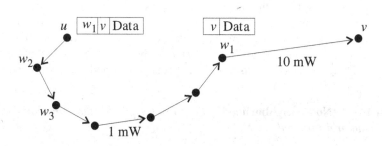

Figure 14.10 Example of TUNNELEDCLUSTERPOW's execution: the intermediate destinations of the packet in the route to the final destination are recursively encapsulated in and decapsulated from the original message.

The software architecture for the TUNNELEDCLUSTERPOW protocol is similar to that of CLUSTERPOW (using multiple routing daemons, and so on). Unfortunately, implementing the recursive encapsulation and decapsulation of packets is a very challenging task, which would require the design of a dynamic per-packet tunneling mechanism. This mechanism is not available in the standard Linux distribution, and it is quite complicated to implement.

There is also another problem with TUNNELEDCLUSTERPOW, that is, the size of the message header, which is considerably increased with respect to the case of CLUSTERPOW. Recursive encapsulation of intermediate node addresses in the packets might entail a notable overhead, especially if the source–destination paths tend to be relatively long.

Because of the issues described above, Kawadia and Kumar decided not to implement TUNNELEDCLUSTERPOW.

14.4   The KNEIGHLEV Protocol

The KNEIGHLEV protocol is a level-based implementation of neighbor-based TC introduced in (Blough et al. 2003c) by the authors of the KNEIGH protocol.

The idea is similar to the one exploited in KNEIGH: connect each node to its k closest neighbors, with the constraint of using only bidirectional wireless links.[4] By setting k properly (and under the assumption of random, uniform node spatial distribution), we have the guarantee that the generated communication graph is connected w.h.p. Also, similar to KNEIGH, nodes change the transmit power level periodically, to maintain a topology with certain features in presence of dynamic network conditions (periodical TC).

However, KNEIGHLEV displays a major difference compared with KNEIGH: its implementation does not require that nodes be capable of estimating their relative distance. Instead, nearest neighbors discovery is done by exchanging control messages at different power levels (see the next subsection). Thus, KNEIGHLEV is based only on the standard working assumption of wireless symmetric medium, and on the assumption that only discrete power level settings $P_1, P_2, \ldots, P_{max}$ are available at each network node.

14.4.1   Protocol description and properties

In KNEIGHLEV specification, nodes can use different transmit power levels $P_1, P_2, \ldots, P_{max}$, which are the same for all the nodes. As a consequence, the communication graph $G = (N, E)$ representing the network topology after each node has set its transmit power level is in general directed, that is, there might exist nodes u, v such that $(u, v) \in E$ (because v is within u's transmitting range), but $(v, u) \notin E$ (because u is out of v's transmitting range). Thus, Blough et al. define three different concepts of node neighborhood, which we report:

Definition 14.4.1 (Node neighborhoods) *Let $G = (N, E)$ be the communication graph, and let u be an arbitrary node in N.*

- *The incoming neighbor set of node u, denoted as $N_i(u)$, is defined as*

$$N_i(u) = \{v \in N : (v, u) \in E\}.$$

---

[4]The reasons for requiring symmetric wireless links are thoroughly discussed in Section 7.4.

– *The outgoing neighbor set of node u, denoted as $N_o(u)$, is defined as*

$$N_o(u) = \{v \in N : (u, v) \in E\}.$$

– *The symmetric neighbor set of node u, denoted as $N_s(u)$, is defined as*

$$N_s(u) = N_i(u) \cap N_o(u) = \{v \in N : (v, u) \in E \text{ and } (u, v) \in E\}.$$

Note that the neighbor sets of node u are modified not only when u changes its own power level but also when nodes in u's neighborhood change their power levels. In particular, node u can modify the size of its N_o set by changing its own transmit power level, while other nodes' power level changes influence the size of u's incoming neighbor set. Thus, the size of the symmetric neighbor set N_s, which must be at least k according to KNEIGHLEV specification, can be modified only by acting on the power levels of both node u and the nodes in its vicinity.

This observation discloses a flaw in some early neighbor-based TC protocols, such as MobileGrid (Liu and Li 2002) and LINT/LILT (Ramanathan and Rosales-Hain 2000). These protocols are based on the idea of maintaining the number of neighbors of each node within a low and a high threshold: if the number of neighbors of a certain node u is too low, u's transmit power is increased; on the contrary, if the number of neighbors is too high, u's transmit power level is decreased. Note that the number of neighbors of a node is estimated by overhearing control and data traffic; that is, what it is estimated is the size of the incoming neighbor set N_i. On the contrary, by acting on its transmit power level, node u can only modify the size of the outgoing neighbor set N_o. So, the controlled parameter (incoming neighbor set size) is not influenced at all by the actions (changing u's transmit power level) taken in response to its incorrect setting.

To circumvent this problem, Blough et al. propose a protocol based on explicit control message exchange between nodes in a neighborhood. Two types of control messages are used: *beacon* and *help* messages. The content of beacon and help messages is the same, that is, ID and current power level of the sender. However, the purpose of beacon and help messages is different. Beacons are used to inform current (outgoing) neighbors of the power level of the sender so that they can compute the size of their symmetric neighbor set. Help messages are used to trigger some nodes in the vicinity to increase their transmit power level so that the number of symmetric neighbors of the help message sender can be increased.

Initially, all nodes set their transmit power to level P_0 and send a beacon message. After sending this initial message and waiting for a stabilization time, node u checks whether it has at least k symmetric neighbors. If so, it becomes *inactive*, and from this point on it participates in the protocol by simply responding (if necessary) to the help messages sent by other nodes. Otherwise, it remains *active*, and it enters the *Increase Symmetric Neighbors* (ISN) phase. During the ISN phase, node u sends help messages at increasing power levels, until the symmetric neighbor set grows to the desired size or the maximum transmit power level is reached.

When node u receives a beacon message $(v, P_i(v))$ from node v (here, $P_i(v)$ denotes the current transmit power level of node v), it first checks whether $v \in N_i(u)$. If so, u has already received a control message from v, and the current beacon is simply ignored. Otherwise, u stores in a local variable $l_v(u)$ the level $P_i(v)$, which represents the minimum power level needed for u to reach node v.[5] Furthermore, node u includes v in its list of incoming neighbors and, in case u's transmit power level is at least $P_i(v)$, also in the symmetric neighbor set.

When node u receives a help message $(v, P_i(v))$ from node v, it checks whether this is the first control message received by v. If so, it sets the $l_v(u)$ variable and the set of incoming and symmetric neighbors as described above. To respond to the help message, node u compares its own power level $P(u)$ with $P_i(v)$ and, in case it is less than $P_i(v)$, it increases its power level to $P_i(v)$. This way, v's symmetric neighbor set will eventually be increased in size by at least one. Furthermore, node u includes v in its symmetric neighbor set.

If the help message $(v, P_i(v))$ is not the first control message received from v, then it must be $v \in N_i(u)$, and node u knows the minimum power level needed to reach v (which is stored in the local variable $l_v(u)$). Thus, node u simply checks whether $v \in N_s(u)$; if so, u is already a symmetric neighbor of v, and the help request from v is ignored. Otherwise, it must be $P(u) < P_i(v)$, and the power level of u is increased to $l_v(u)$ (which is the minimum level needed to make u and v symmetric neighbors).

Note that particular attention must be paid when increasing the power level of node u from its current value to $P_i(v)$ (or to $l_v(u)$). In fact, we must guarantee the property that when the variable $l_u(y)$ is set at node y, it actually stores the minimum power required for y to reach node u. So, when increasing its power level, node u sends a sequence of beacons, one at each power level from its current power level to the target power level.

The KNEIGHLEV protocol is summarized in Figure 14.11. In order to improve readability, we drop the 'argument' u (which is clearly redundant at node u) from the variables $l_x(u)$, $N_s(u)$, and $N_i(u)$.

In (Blough et al. 2003c), Blough et al. show the following:

- *Message complexity*: KNEIGHLEV exchanges at most $O(nh)$ overall (h is the number of power levels) to compute the final topology.

- *Symmetry*: KNEIGHLEV preserves symmetry, that is, at the end of its execution, $u \in N_s(v)$ if and only if $v \in N_s(u)$.

- *Connectivity*: The topology built by KNEIGHLEV is connected w.h.p., under the assumption that k is appropriately chosen (see Section 14.4.4) and that the nodes are distributed uniformly at random in the unit square.

- *Minimal power increase*: Node u sends the help message at power level P_i if and only if the number of nodes in u's transmitting range at power level P_{i-1} is smaller than k;

- *Symmetric neighbor count*: At the end of KNEIGHLEV's execution, a node has at least k symmetric neighbors, or it transmits at maximum power.

---

[5]Here, the assumptions of symmetric wireless medium and of all the nodes using the same power levels are essential for KNEIGHLEV's correctness.

Algorithm KNEIGHLEV:
(algorithm for node u)

P_i is the i-th transmit power level
N_i, N_o and N_s are the incoming, outgoing, and symmetric neighbor sets of node u
$P(u)$ is the current transmit power level of node u
k is the target number of symmetric neighbors (input parameter)

1. Initialization
    $P(u) = P_0$
    $N_i = N_o = N_s = \emptyset$
    send beacon $(u, P(u))$
    wait for stabilization time

2. Checking the symmetric neighbors count
    repeat until $(|N_s| \geq k)$ or $(P(u) = P_{max})$
        set $P(u)$ to the next higher power level
        for any $z \in N_i$ such that $P(u) = l_z$ do $N_s = N_s \cup \{z\}$
        send help $(u, P(u))$
        wait for stabilization time
    $P(u)$ is the final transmit power level of node u

3. Message handling
3a. Upon receiving a beacon message $(v, P_i(v))$
    if $v \notin N_i$
    $l_v = P_i(v)$
    $N_i = N_i \cup \{v\}$
    if $P(u) \geq P_i(v)$ then $N_s = N_s \cup \{v\}$
3b. Upon receiving a help message $(v, P_i(v))$
    if $v \in N_i$ and $v \notin N_s$ then $StepwiseIncrease(P(u) + 1, l_v)$
    if $v \notin N_i$ then
    $l_v = P_i(v)$
    $N_i = N_i \cup \{v\}$
    if $P(u) < l_v$ then $StepwiseIncrease(P(u) + 1, l_v)$
    else $N_s = N_s \cup \{v\}$

4. Procedure $StepwiseIncrease(P_h, P_j)$
    for $t = h, \ldots, j$ do
    set $P(u) = P_t$
    send beacon $(u, P(u))$
    for any $z \in N_i$ such that $P(u) = l_z$ do $N_s = N_s \cup \{z\}$

Figure 14.11 The KNEIGHLEV protocol.

14.4.2   Optimizations: the KNEIGHLEVU protocol

As the examples reported in this section show, the KNEIGHLEV protocol leaves space for some optimizations.

A first simple optimization is along the guidelines of the shrink-back optimization proposed for the CBTC protocol (see Section 11.1): suppose there exists a node u having less than k symmetric neighbors when transmitting at maximum power; this node can reduce its transmit power to lower levels as long as this does not reduce its symmetric neighbors count.

A second, more serious opportunity for optimization is motivated by the example reported in Figure 14.12: node u's current power level is P_2, resulting in covering all the nodes in area A_2. Thus, this transmit power is sufficient to have at least $k - 1$ symmetric neighbors. According to KNEIGHLEV's specification, u then sends a help message at power level P_3, forcing *all* the nodes in area $A_3 - A_2$ to increase their power to level at least P_3. This mechanism induces unnecessary power increases, because all the four nodes in $A_3 - A_2$ must increase their power level, while the power increase of a single node would be sufficient to meet the requirement on the symmetric neighbors count at node u.

A more subtle situation in which KNEIGHLEV displays inefficient behavior is depicted in Figure 14.13. The target number of symmetric neighbors is $k = 4$. At the current power level P_1, node u has at most three symmetric neighbors. To increase the symmetric neighbors count, node u sends a help message at power level P_2, forcing node v, which currently has power level P_0, to increase its power to level P_2. Then, the overall power increase for node u to have the target number of symmetric neighbors is $P_2 - P_1$ for node u, and $P_2 - P_0$ for node v. Note that there exists a third node in this scenario, node w, whose power level

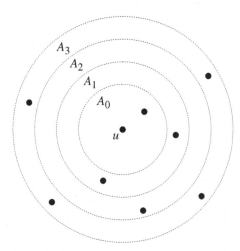

Figure 14.12 Example of inefficient KNEIGHLEV's behavior. The target number of neighbors is $k = 5$. Area A_i denotes the radio coverage of node u at power level P_i. At power level P_2, node u has at most $4 = k - 1$ symmetric neighbors. Thus, it sends the help message at power level P_3, forcing all the nodes in $A_3 - A_2$ to increase their power to level at least P_3.

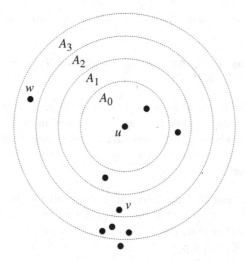

Figure 14.13 A second, more subtle example of inefficient KNEIGHLEV's behavior. The target number of neighbors is $k = 4$. The power level of node v is P_0, and the power level of node w is P_3. At power level P_1, node u has at most $3 = k - 1$ symmetric neighbors. Instead of sending the help message at power level P_2, node u might autonomously increase its power level to P_3, thus fulfilling the symmetric neighbor count requirement without forcing node v to increase its power level.

is set to P_3, and thus it is in u's incoming neighbor set. Node u is aware of node w and of its current power level. So, instead of sending the help message at power level P_2, it could decide to autonomously increase its power level to P_3. With this power level setting, node u would satisfy the symmetric neighbor count requirement. Note that this second option is nonoptimal from node u's point of view (it increases u's power level of two steps, instead of a single step as in the standard KNEIGHLEV specification); however, this solution might be optimal from a local neighborhood point of view: if

$$(P_2 - P_1) + (P_2 - P_0) > (P_3 - P_1),$$

it is locally optimal to increase u's transmit power of two steps. For instance, if $P_0 = 1$ mW, $P_1 = 5$ mW, $P_2 = 20$ mW and $P_3 = 30$ mW, the 'unselfish' behavior of node u (the second option described above) is locally optimal from the energy consumption point of view. Conversely, if $P_0 = 5$ mW, $P_1 = 20$ mW, $P_2 = 30$ mW and $P_3 = 50$ mW, the situation is reversed.

The examples shown in Figures 14.12 and 14.13 motivated the authors of (Blough et al. 2003c) to design an optimized version of KNEIGHLEV, which is called KNEIGHLEVU (U stands for 'unselfish').

Three additional types of control messages are used in KNEIGHLEVU: *enquiry*, *reply*, and *selective help* messages. Enquiry and reply messages contain the same information as beacon and help messages (sender ID and transmit power level), while the selective help message contains the ID of the sender and the ID(s) of the node(s) that must increase its (their) transmit power levels.

Suppose node u, which currently has less than k symmetric neighbors, is transmitting at power level P_i:

1. Instead of sending a help message at power level P_{i+1}, node u sends an enquiry message at power level P_{i+1}.

2. When a node (say, node v) in u's transmitting range whose current power level is below P_i receives the enquiry message, it does not immediately step up the power level; instead, it sends to u a reply message at *temporary* power level P_i. The purpose of this message is to inform u that v is a potential helper, which is currently using a certain power level.

3. Once node u has gathered the information from all its potential helpers, it is able to compute the locally optimal solution, accounting also for the nodes in $N_i(u)$ that might become symmetric neighbors by simply increasing u's power level. Then, node u schedules one of several possible actions, depending on the locally optimal solution that has been identified. In particular, u can

   (a) increase its own power level, so as to reach more nodes in $N_i(u)$;

   (b) send a selective help message, asking some of the nodes in the neighborhood to increase their power levels;

   (c) send a generalized (i.e. old style) help message.

Observe that KNEIGHLEVU might require some nodes to perform temporary power level increases (when sending reply messages), thus partially impairing KNEIGHLEV's philosophy of periodical TC. However, contrary to the case of per-packet TC, these temporary power level changes occur only during the network setup phase, and on an occasional basis.

We remark that the local optimization performed by KNEIGHLEVU is only a *heuristic* used to maintain the power consumption as low as possible: in fact, the composition of locally optimal configuration might result in a globally nonoptimal solution. Furthermore, a solution that is locally optimal at a certain time might become suboptimal later on (e.g. because a certain node in u's neighborhood would have stepped up its power level anyway to increase the symmetric neighbors count of another node).

14.4.3   KNEIGHLEV versus KNEIGHLEVU

It interesting to compare the relative performance of KNEIGHLEV and KNEIGHLEVU. In general, KNEIGHLEVU allows a finer tuning of the symmetric neighbors count, which should result in better energy efficiency. On the other hand, KNEIGHLEVU tends to exchange more control messages than KNEIGHLEV does because of the up to three phase interaction (enquiry–reply–help) between neighbor nodes. So, it is not clear which one of the two protocols has the best overall performance.

To investigate this point, Blough et al. have performed a set of simulation experiments in which the protocols' performance is evaluated according to the following metrics:

– *energy cost*, which is defined as the sum of the power levels of all the network nodes at the end of the protocol execution;

- *average logical and physical node degrees*, which are defined as in Section 9.3;

- *message overhead*, defined as the average number of messages sent per node during the protocol execution.

The simulation results reported in (Blough et al. 2003c) are obtained considering a square simulation area in which a certain number of nodes is distributed uniformly at random. Nodes can use six different power levels, which are taken from the specifics of the CISCO Aironet 350 wireless card: 1, 5, 20, 30, 50, and 100 mW. The simulation results have shown the following:

- KNEIGHLEVU performs better than KNEIGHLEV with respect to both energy cost and average node degree (both logical and physical). The relative advantage of KNEIGH-LEVU with respect to KNEIGHLEV in terms of energy cost is considerable (it can be as high as 50% for networks of large size $-n = 500$), while the relative advantage in terms of average node degree is more limited (in the order of 15%).

- In terms of message overhead, KNEIGHLEV exchanges relatively less messages than KNEIGHLEVU when the size of the network is small to medium ($n = 300$ and below), while the situation is reversed for large networks (n above 300). However, in all the network sizes considered, the relative difference between the number of messages exchanged by the two protocols is quite small (in the order of 20%). In absolute terms, a node sends an average of 6 (when $n = 50$) to 3 (when $n = 500$) messages with KNEIGHLEV, and an average of 9 (when $n = 50$) to 2.8 (when $n = 500$) messages with KNEIGHLEVU.

Blough et al. have also compared the performance of their protocols to that of a level-based implementation of CBTC (see Section 11.1 for a description of the CBTC protocol). In the level-based implementation of CBTC, the final transmit power of a node as computed by CBTC is rounded up to the next higher power level available. The simulation results have shown that KNEIGHLEV and KNEIGHLEVU performs slightly better than level-based CBTC with respect to both energy cost and average node degree. However, it must be observed that CBTC preserves network connectivity in the worst case, while KNEIGHLEV and KNEIGHLEVU give only a probabilistic guarantee on network connectivity.

The authors of (Blough et al. 2003c) consider also the issue of optimizing the choice of the transmit power levels: instead of using the predefined levels of the CISCO Aironet 350 card, one might think of selecting the power levels according to some criterion. Blough et al. consider a scenario in which nodes are distributed uniformly at random in a square region, and set the power levels in such a way that the expected number of neighbors at every power level is evenly distributed (which does not occur with the power levels set as in the specifics of the CISCO wireless card). With this optimization implemented, the energy cost of the topology generated by KNEIGHLEVU is reduced further (in the order of 10%). Although the achieved energy saving is quite limited, it is interesting to note that *by simply acting on the choice of the power levels, one might reduce energy consumption.*

14.4.4   Setting the value of k

Similar to the case of KNEIGH, the performance of the KNEIGHLEV and KNEIGHLEVU protocols heavily depends on the choice of the target number k of symmetric neighbors:

choosing a value of k that is too small leads to the formation of a disconnected network topology, while setting k to a value that is too large is detrimental for energy efficiency and spatial reuse. So, the problem is that of identifying the minimal value of k that achieves network connectivity, that is, the critical neighbor number (CNN).

Observe that the problem of determining the CNN has been thoroughly discussed in 12.1. However, in that case, the value k upper bounded the number of *outgoing* neighbors of a node, some of which might be nonsymmetric. On the contrary, KNEIGHLEV and KNEIGHLEVU try to maintain the number of *symmetric* neighbors of a node equal to (or above) k. Furthermore, KNEIGHLEV and KNEIGHLEVU are level-based protocols,[6] so in many cases it might be infeasible to obtain exactly k symmetric neighbors.

Despite the differences outlined above, Theorem 12.1.7 presented in Section 12.1 can be used to prove the following result (see (Blough et al. 2003c)):

Theorem 14.4.2 (Blough et al. 2003c) *Let n nodes be placed uniformly at random in $R = [0, 1]^2$, and assume that each node in the network has at least k other nodes in its transmitting range when transmitting at maximum power. Let G^k_{KNL} be the network topology that is built when KNEIGHLEV is executed with parameter k. If $k \in \Theta(\log n)$, then G^k_{KNL} is connected a.a.s.*

Blough et al. prove that a similar result also holds for the communication graph generated by KNEIGHLEVU.

Note that the result stated in Theorem 14.4.2 is weaker than the analogous theorem that holds for KNEIGH, since it only proves that $k \in \Theta(\log n)$ neighbors are sufficient (but they might not be necessary) for connectivity.

Blough et al. give also a more practical, simulation-based evaluation of the desired number of symmetric neighbors. They first consider the KNEIGHLEV protocol, and show that a value of k equal to 4 is sufficient to generate well connected topologies (at least 98% of the generated graphs are fully connected) for networks of size 150 and above. For smaller networks, slightly larger values of k are needed to achieve the same degree of network connectivity (we have $k = 6$ when $n = 50$). Then, Blough et al. analyzed the topology generated by KNEIGHLEVU, and they verified that if the same value of k as in KNEIGHLEV is used, KNEIGHLEVU generates relatively less connected topologies: on the average, 90% of the generated graphs are fully connected and, in case the generated graph is disconnected, at least 95% of the network nodes are in the same connected component.

14.5   Comparing CLUSTERPOW and KNEIGHLEV

Since CLUSTERPOW and KNEIGHLEV are the only existing level-based approaches to TC that use heterogeneous power assignments, it is interesting to perform a qualitative comparison of their main features.

A first major difference between CLUSTERPOW and KNEIGHLEV is from an architectural point of view (see Figure 14.14). CLUSTERPOW is a protocol for jointly performing (implicit) node clustering, routing, and transmit power control. So, CLUSTERPOW can be seen as a network layer protocol, which interacts both with the transport and the MAC

---

[6]We recall that in the CNN analysis of Section 12.1 it is implicitly assumed that each node in the network can set its transmit power to a level such that it reaches exactly k other nodes.

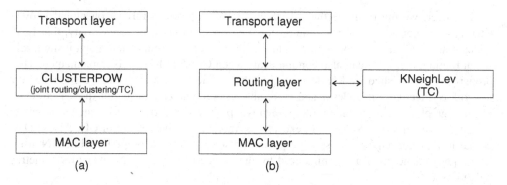

Figure 14.14 The CLUSTERPOW (a) and KNEIGHLEV (b) protocols in the protocol stack.

layer. Although not explicitly stated in (Blough et al. 2003c), KNEIGHLEV is based on a different architectural view: this protocol is a separate module that is executed periodically, and sets the power levels that nodes will use for route discovery and maintenance. Thus, KNEIGHLEV can be seen as a network layer module, which, however, is separated from the routing module.

This different view on where the TC functionality should be located in the protocol stack has led to different design choices in the two approaches: nodes in CLUSTERPOW change the power level on a per-packet basis (this is due to the fact that CLUSTERPOW performs routing and transmit power control at the same time), while KNEIGHLEV is based on periodical power level changes. As discussed in Chapter 13, periodical power level settings are in general preferable in mobile or highly dynamic networks. On the other hand, per-packet TC has the potential of achieving better benefits in terms of reduced energy consumption and increased spatial reuse with respect to periodical TC (provided the technological problems described in Section 14.3.2 can be solved).

Another consequence of CLUSTERPOW designers' choice of integrating routing and power control in the same module is the fact that CLUSTERPOW is based on global knowledge: when node u has to send a packet to a faraway node v, it must know the minimum power level needed to reach v, an information that can be obtained only by circulating a certain number of control messages in the network. As discussed in detail in Chapter 9, networkwide information exchange should be avoided in TC protocols, as this, in general, induces a high control message overhead for generating the desired topology. This is actually the case with CLUSTERPOW, which causes a considerable increase in the routing overhead with respect to traditional routing (with no integrated TC functionality): if nodes can set the transmit power level at h different levels, the routing overhead with CLUSTERPOW is increased by a factor h with respect to the case of no TC.

Contrary to CLUSTERPOW, KNEIGHLEV is a localized protocol: a node exchanges messages only with its immediate neighbors in the maxpower communication graph. In turn, this implies a low control message overhead as compared to CLUSTERPOW: when the KNEIGHLEV is invoked (we recall that in this approach the TC protocol is reexecuted periodically), each node sends only few messages to immediate neighbors (typically, less than six) before setting the transmit power to the appropriate level.

However, we must outline that CLUSTERPOW's choice of relying on global knowledge has a major advantage also: the protocol ensures full network connectivity, that is, if nodes u and v are able to communicate in the maxpower communication graph using traditional routing, then they can also communicate when CLUSTERPOW is implemented. This property is not satisfied by KNEIGHLEV, which provides only a probabilistic guarantee on network connectivity: if nodes u and v are able to communicate in the maxpower communication graph using traditional routing, then w.h.p. they can also communicate in the graph obtained at the end of KNEIGHLEV' execution. Another positive feature of CLUSTERPOW is that it does not require any specific parameter setting to work properly, while KNEIGHLEV's performance is heavily influenced by the choice of the target number of symmetric neighbors k.

A comparison of the CLUSTERPOW's and KNEIGHLEV's performance in terms of energy efficiency and increased spatial reuse is not immediately evident: as a rule of thumb, we can say that when sending a packet to a faraway node CLUSTERPOW is more efficient than KNEIGHLEV in the downlink, while the situation is reversed in the uplink.

This point is explained in Figure 14.15. Suppose node u wants to send a packet to node v. There are three levels (1, 10, and 100 mW), which, in CLUSTERPOW, defines a cluster hierarchy with three levels. Since nodes u and v are in the same 100 mW cluster, but in different 10 mW clusters, the packet is initially sent at 100 mW, and it is relayed to the destination using increasingly lower power as it gets closer to the destination. The total power consumption for delivering the packet with CLUSTERPOW is 211 mW. Let us now consider the case of KNEIGHLEV. Assume that the power level of nodes u, z_4, w_1 and w_2 at the end of the protocol execution are set as follows: 1, 10, 100, and 100 mW, respectively. So, sending the packet from u to v consumes 211 mW overall in the case of KNEIGHLEV also. However, we note that KNEIGHLEV is more efficient than CLUSTERPOW in the

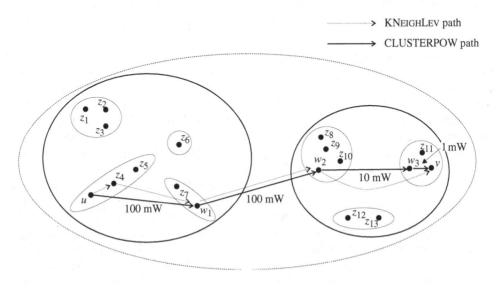

Figure 14.15 Example showing the relative energy and interference performance of the CLUSTERPOW and KNEIGHLEV protocols.

Table 14.1   A qualitative comparison of the CLUSTERPOW and
KNEIGHLEV protocols

CLUSTERPOW	KNEIGHLEV
Joint clustering/routing/TC	TC is a separate net layer module
Per-packet TC	Periodical TC
Global knowledge required	Localized
High control overhead	Low control overhead
Guarantees full connectivity	Guarantees connectivity w.h.p.
No parameter setting	Setting k is required
Efficient in the downlink	Efficient in the uplink

uplink (from node u to the intermediate node w_1), while the situation is reversed in the downlink (from node w_1 to node v).

We remark that the situation depicted in Figure 14.15 is only an example and, depending on the node configuration and on the traffic pattern, situations can be easily identified in which CLUSTERPOW is more efficient than KNEIGHLEV, or the reverse holds.

The qualitative comparison of CLUSTERPOW and KNEIGHLEV discussed in this section is summarized in Table 14.1.

15

Open Issues

In this book, we have presented several approaches to the topology control problems that have appeared in the literature. Although the body of research devoted to this topic is considerable, many aspects related to the application of TC techniques in ad hoc and sensor networks have not been carefully investigated yet. In this chapter, we discuss the most important such aspects, indicating several avenues for further research in the field.

15.1 TC for Interference

In Chapter 3, we have discussed the potential advantages of using TC techniques in wireless ad hoc and sensor networks: reducing node energy consumption and reducing radio interference between nodes. Despite the acknowledged motivations for using TC being twofold, the vast majority of research on TC focused only on one of them, that is, on reducing node energy consumption. Thus, natural questions such as What network topologies turn out to be good from the radio interference point of view? Are these topologies the same, similar, or very different from the 'energy-efficient' topologies studied so far? remain to be answered.

Answering these questions is relatively simple if all the nodes are assumed to use the same transmit power level (homogeneous TC). In fact, it can be seen that setting the transmitting range to the critical value for connectivity is the best solution for reducing radio interference (this is formally proven in (Narayanaswamy et al. 2002) – see Section 14.2). Thus, setting the common transmitting range to the critical value for connectivity turns out to be optimal for reducing both node energy consumption and radio interference.

The situation is more complicated if nodes are allowed to use different transmit power levels. The issues of determining interference-optimal topologies and investigating the differences/similarities between energy-optimal and interference-optimal topologies have not been addressed in the current literature. Only very recently, a few papers have tackled the above issues.

Burkhart et al. (2004) introduce the notion of *edge coverage* to model radio interference in ad hoc/sensor networks. Edge coverage is formally defined as follows:

Topology Control in Wireless Ad Hoc and Sensor Networks   P. Santi
© 2005 John Wiley & Sons, Ltd

Definition 15.1.1 (Edge coverage) *Let $e = (u, v)$ be any edge of the maxpower graph $G = (N, E)$, indicating that nodes $u, v \in N$ are within each other's maximum transmitting range. The coverage of edge e is defined as*

$$Cov(e) = |\{w \in N : w \text{ is inside } D(u, \delta(u, v))\} \cup$$

$$\{w \in N : w \text{ is inside } D(v, \delta(u, v))\}|,$$

where $D(x, y)$ denotes the disk of radius y centered at node x and $\delta(x, y)$ is the distance between nodes x and y.

An example clarifying the notion of edge coverage is reported in Figure 15.1.

The notion of edge coverage can be used to estimate the expected interference experienced when communicating along a certain link, at least when the MAC layer is based on CSMA-CA (as it is the case of IEEE 802.11, and of the vast majority of MAC protocols for sensor networks). In fact, with this implementation of the MAC layer, nodes u and v exchange RTS/CTS messages before performing the communication, and all the nodes receiving either one of the RTS/CTS messages (i.e. at least all the nodes in $D(u, \delta(u, v)) \cup D(v, \delta(u, v))$) refrain from their communications to avoid interference with the transmission along link (u, v). So, the number of nodes in $D(u, \delta(u, v)) \cup D(v, \delta(u, v))$ is a measure of the amount of wireless medium 'consumed' by the transmission, that is, of the interference generated by the communication along link (u, v).

Note that edge coverage in general tends to underestimate the generated interference because of two reasons: (i) the possible use of nonminimal transmit power levels and (ii) carrier sensing range being larger than the transmitting range.

As for (i), we observe that the definition of edge coverage relies on the assumption that the minimum possible power level is used by both node u and node v when communicating along link (u, v). Because of technological (e.g. only a limited number of different transmit power levels available) and/or design (e.g. use of periodical TC techniques) constraints, using this minimum power level might not always be possible.

Regarding issue (ii), we observe that in many MAC protocol implementations (e.g. IEEE 802.11) a carrier sensing mechanism is used in combination with RTS/CTS message

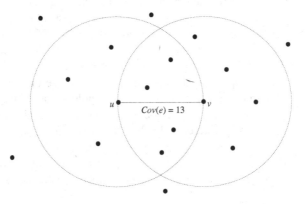

Figure 15.1 Coverage of edge (u, v).

exchange to regulate the access to the channel: if the sender node detects a carrier on the channel, it does not even send the RTS message, and it refrains from packet transmission until no carrier is detected. In general, the carrier sensing range of a certain transmission is larger than the transmitting range. For instance, in the IEEE 802.11 standard, it is typically assumed that the carrier sensing range is about twice as long as the transmitting range. A consequence of this observation is that not only do the nodes in $D(u, \delta(u, v)) \cup D(v, \delta(u, v))$ refrain from their communications when nodes u and v are communicating but also the nodes in the carrier sensing range of u and v are likely to detect a carrier and refrain from sending packets.

Despite issues (i) and (ii), edge coverage is still a reasonable measure of the expected node interference, and it has been used by Burkhart et al. in (Burkhart et al. 2004) to define the notion of *interference* of a communication graph, which is defined as follows:

Definition 15.1.2 (Graph interference) *Let $G = (N, E)$ be any communication graph. The interference of graph G is the maximum coverage over all possible edges in E. Formally,*

$$I(G) = \max_{e \in E} Cov(e).$$

The authors of (Burkhart et al. 2004) consider the problem of finding a connected subgraph of the maxpower graph with minimum interference and provide a simple centralized algorithm that computes the optimal solution to this problem. It is easy to see that the interference-optimal topology in this context is essentially the MST built on the maxpower graph, where edges are weighted with their coverage. Furthermore, Burkhart et al. show that a topology in which every node establishes a symmetric connection to at least its closest neighbor (as it is the case of all the topologies built by the TC protocols for reducing energy consumption presented in this book) can be $\Omega(n)$ times away from the interference-optimal topology. In other words, they show that, at least in a worst-case scenario, *reducing node energy consumption and reducing radio interference are conflicting goals*. This observation is coherent with a theoretical result presented in (MeyerAufDerHeide et al. 2002), which, although based on a more complicated notion of interference, states that finding a topology that reduces node energy consumption and finding an interference-optimal topology are conflicting goals.

Burkhart et al. also consider the problem of finding an interference-optimal topology with the additional requirement of building a distance t-spanner of the maxpower graph and present centralized and localized algorithms to solve this problem.

The problem of building low-interference topologies has been addressed in (Moaveni-Nejad and Li 2005) also. In particular, Moaveni and Li consider different measures of interference of a graph and provide an algorithm for computing the interference-optimal, connectivity-preserving topologies according to the defined metrics. For instance, one of the interference metrics considered is the *average* of the edge coverage in a graph, and Moaveni and Li show that the interference-based MST (i.e. the same topology defined in (Burkhart et al. 2004)) is optimal with respect to this metric also.

Although some initial steps toward the design of interference-efficient topologies have been recently done, we believe a lot of research on this topic is still to be performed. In particular, we cite two shortcomings of the above cited approaches to interference-optimal TC: (i) reliance on a specific radio channel model and (ii) disregarding the effect of multihop communications.

With respect to (i), we observe that all the interference metrics proposed so far are based on the notion of edge coverage introduced in (Burkhart et al. 2004). Unfortunately, edge coverage is based on a purely geometric notion (number of nodes in the intersection of two disks), and it implicitly relies on the assumption that the radio coverage area is a perfect circle. As we have discussed in Chapter 2, this assumption is widely accepted in the wireless ad hoc network research community, but it is unlikely to hold in most practical situations. Thus, the definition of an interference measure that can be used in combination with more general radio channel models (such as the log-normal shadowing model) is an urgent need.

Shortcoming (ii) can be even more serious, as the example reported in Figure 15.2 illustrates. In the figure, edges are labeled with their coverage, which is used to build the MST, which is represented by the bold edges (this is the topology computed by the interference-optimal algorithm presented in (Burkhart et al. 2004)). Although the MST minimizes the maximum possible interference level on a single link, it turns out to be a very bad topology from the multihop interference point of view: the path connecting nodes u and v in the MST with minimum interference cost (calculated by summing up the coverage of the edges in the path) has cost equal to 24; however, the two nodes can be connected through a single link (dashed edge) in the maxpower graph at a cost of 6, which is four times lower than the cost of the optimal (u, v) path in the 'supposed to be' interference-optimal topology. A similar observation applies to all the graph interference measures introduced in the literature so far. Thus, what is still needed is the definition of a different notion of graph interference that accounts for interference in multihop communications (we remark that this is the very distinguishing feature of ad hoc and sensor networks!). On the basis of this notion, topologies that perform well with respect to multihop interference must be investigated.

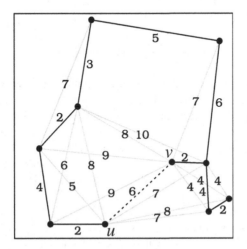

Figure 15.2 Example showing that the interference-based MST is not a good topology when you account for multihop interference.

15.2 More-realistic Models

In Chapter 2, we have described the models used in the wireless ad hoc network community to model the radio link and node energy consumption. As discussed therein, the issue in the definition of the network model is to find a good balance between *simplicity* and *representativeness* of the model: on one hand, it must be kept simple enough to allow the derivation of meaningful theoretical results; on the other hand, it must be accurate enough to ensure that the findings obtained by using the model are useful in practical situations.

We believe that finding such a good balance in the realm of wireless ad hoc networks will be a very difficult task and that models achieving this good balance have not been identified yet. In fact, both the radio channel and the energy model so far used in the analysis of ad hoc network properties are probably too simplistic and inaccurate. In the following, we discuss in detail the issues of radio channel and energy modeling.

15.2.1 More-realistic radio channel models

The radio channel model that is typically used in the analysis of wireless ad hoc network properties is based on the assumption of homogeneous decay with distance of the radio signal (i.e. circular radio coverage area). This corresponds to using the log-distance path model for the radio channel (see Section 2.1), and the communication graph built according to this model is essentially a point graph (a.k.a. unit disk graph or a geometric random graph).

Thanks to the fact that unit disk graphs and geometric random graphs have been analyzed in depth in the applied probability community (see Appendix B), the assumption of homogeneous radio signal decay has allowed the derivation of a considerable body of theoretical results concerning fundamental properties of ad hoc networks. However, it is widely known that the above mentioned assumption of circular radio coverage area is a sufficiently accurate model of the actual radio signal propagation only in very specific cases, such as networks deployed in open air, flat environments. Unfortunately, in real-world applications, it is likely that ad hoc and sensor networks are used in different environments: typically, ad hoc networks are deployed in urban scenarios, where the presence of buildings, obstacles, and so on renders the propagation of the radio signal in the air highly irregular; the same happens in typical applications of WSNs, which are either indoor (e.g. intrusion detection in a building) or in harsh environments.

On the basis of the above discussion, there is the strong feeling that the many characterizations of ad hoc network properties so far derived in the literature turn out to be scarcely useful in practice and that a radio model that accounts for shadowing/fading effects should be used instead of the simplistic log-distance path model.

In Section 4.4, we have presented a set of theoretical and experimental results that can be regarded as initial steps in this direction. We also want to mention the *Random Vertex* model recently proposed by Faragó (Faragó 2002), in which the link between nodes u and v occurs with probability $p_l(u, v)$, where $p_l(u, v)$ is an arbitrary function (with values in $[0, 1]$) with the property that $\delta(u, v) \leq \delta(u, w)$ implies $p_l(u, v) \geq p_l(u, w)$. Furthermore, Faragó introduces the concept of *node availability*, that is, node u is available with a given probability $p_a(u)$ (this is similar to the model of Bernoulli nodes – see Section 6.2). Faragò shows that the study of monotone properties[1] (e.g. connectivity) of certain Random Vertex graphs can be reduced to the study of the same properties of a properly defined random graph. This result is potentially interesting, given the overwhelming literature on traditional random graphs (Bollobás 1985; Palmer 1985). Whether the Random Vertex model could be used to derive analytical characterizations of fundamental network properties (e.g. the critical transmitting range for connectivity, k-connectivity, and so on) in a more realistic setting is an open question.

Another possibility to make the radio channel model more realistic is to take into account interference between nodes. For example, in (Dousse et al. 2003), Dousse et al. propose a model in which a bidirectional link between nodes u and v exists if the SINR at the receiver is larger than some threshold, where the noise is the sum of the contribution of interferences from all other nodes and a background noise. The authors analyze the impact of such wireless link model on network connectivity.

Although some research on the characterization of fundamental network properties with a more-realistic radio link model has been recently done, further investigation in this direction is needed.

15.2.2 More-realistic energy models

We have observed in Section 15.1 that most of the literature on TC focused on the problem of defining energy-efficient network topologies. However, the energy model commonly used in the literature accounts only for one component of the energy that is actually consumed in a wireless network card. In fact, as discussed in Section 2.3, the power consumption of a radio interface is determined by at least two main components: the power needed to amplify the transmitted signal (also called the *radiated power*, or transmit power), and the power consumed by the other circuitry of the radio interface (RF/IF converter, IF modem, and so on), which does not depend on the intensity of the radiated signal.

Current TC literature considers only the first component of the power consumption in a real wireless card, the radiated power. Since the radiated power needed to transmit at distance d is proportional to $\frac{1}{d}$ raised to a certain power (the path loss exponent), researchers have argued that from the energy consumption point of view several short hops are better than one long hop (see Section 3.1.1). Unfortunately, this conclusion might be false if a more-realistic energy model, which accounts also for the power consumed in the other components of the radio interface, is used. In fact, as several measurements performed on off-the-shelf wireless cards have shown, the power consumed by the other circuitry of the radio interface is at least as high as, if not higher than, the power consumed by the RF amplifier. Thus, *the constant power consumption involved in any transmission performed by a network node cannot be disregarded*, as it is commonly done in current TC literature.

Furthermore, we should consider that in a radio communication a nonnegligible amount of energy is consumed at the receiver node also to receive and decode the transmitted signal. *This contribution to the overall energy consumption is also disregarded in current TC literature.*

---

[1]A property P of a graph is *monotone* whenever $P(G) \Rightarrow P(G')$, where G' is a supergraph of graph G and $P(G)$ (respectively, $P(G')$) denotes that graph G (respectively, G') satisfies P.

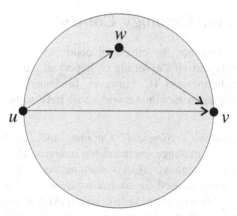

Figure 15.3   Example showing that using a more realistic energy model might lead to radically different conclusions about what an energy-efficient network topology is.

We believe that the definition of a more-realistic energy model may be fundamental in the design of TC protocols as radically different conclusions about 'what an energy-efficient topology' is might be drawn if the constant terms in the node power consumption are taken into account. The following example supports our feeling.

Consider the typical situation depicted in Figure 15.3, which is essentially the same as Figure 3.1: node u wants to send a packet to node v, and we must decide whether it is more energy efficient to send the packet directly to v or to use the intermediate node w to relay the packet. If we disregard the energy consumed by circuitry other than the RF amplifier at the sender end, and the energy consumed at the receiver end, we can conclude that relaying through w is more energy efficient than the direct transmission whenever w lays in the shaded area (see Section 3.1.1).

Let us now include the energy consumed by circuitry other than the RF amplifier at node u in this analysis, and let us consider the power measurements performed on a real wireless card, such as those referring to a CISCO Aironet PC4800 card reported in (Ebert et al. 2002). With this energy model, a node consumes 1.9 W when transmitting at maximum power (corresponding to a nominal transmit power of 50 mW), and 1.48 W when transmitting at minimum power (nominal transmit power 1 mW). So, under the assumption that v is within u's maximum transmitting range (otherwise the direct communication between u and v would be impossible), we have that sending the packet directly has a power cost of at most 1.9 W. On the other hand, relaying the packet through w would require two transmissions, each consuming at least 1.48 W. Since $1.9 < 2 \cdot 1.48 = 2.96$, we can conclude that with this more-realistic energy model sending the packet directly to node v is always the best solution! Note that in this analysis we have not considered the power consumed at the receiver end of the communication, which would also play in favor of the single-hop transmission.

The example above has clearly shown that the energy model has a strong influence on the choice of the energy-efficient links and that accounting for only the radiated power is likely to lead to incorrect conclusions about what an energy-efficient topology is. Thus, further research along this line is urgently needed.

15.3   Mobility and Topology Control

In Chapter 13, we have discussed the effects of node mobility on the design and implementation of TC protocols and on the setting of important parameters such as the critical neighbor number in neighbor-based TC. However, fundamental questions concerning the application of TC techniques in mobile networks still have to be answered. In particular, we cite the following:

1. *Is mobility beneficial or detrimental?* On one hand, we have seen that mobility results in an increased message overhead for maintaining the desired topology (see Chapter 13). This has a negative effect on both network capacity (because a portion of the available bandwidth is used for control messages) and node energy consumption (because nodes send/receive more control messages). On the other hand, mobility has also positive effects on both capacity and energy consumption. In fact, it is known that node mobility can be seen as a means to increase network capacity, provided the delay in packet delivery is not a primary concern (Grossglauser and Tse 2001). As for energy, mobility has the positive effect of balancing the node power consumption: in stationary networks, if a node u uses twice the transmit power of another node v, it is likely to deplete its battery much faster than node v. In presence of mobility, nodes change the transmit power dynamically, and a more balanced energy consumption (with positive effects on network lifetime) is likely to occur. Given this picture, it is still not clear what the overall effect of node mobility is on the network capacity increase and lifetime extension potentially achieved by TC mechanisms.

2. *Determination of the optimal frequency for reconfiguration*: As discussed in Section 13.2, in presence of node mobility, there is a clear trade-off between the message overhead generated by the repeated execution of a TC protocol and the quality of the constructed topology: the more frequently the protocol is reexecuted (i.e. the higher the message overhead), the higher the quality of the constructed topology (e.g. a topology that preserves connectivity). A careful investigation of this trade-off, which would help in answering the issue (1), is still lacking in the literature.

15.4   Considering MultiHop Data Traffic

As witnessed by the considerable body of research reported in this book, a great deal of attention has been devoted to the identification of efficient network topologies, with a particular emphasis on energy efficiency. However, a common approach in the literature is to consider the TC problem as a stand-alone problem, which is analyzed and solved under a graph-theoretic perspective. This type of approach has resulted in the message conveyed by current TC literature: *the sparser the network topology, the better* (provided certain spanning properties of the graph are satisfied).

Indeed, we believe that the problem of determining the 'optimal' network topology cannot be solved by assigning a weight to the links and building a spanner graph, as it is commonly done in current TC literature: in fact, what is an optimal topology depends on several factors, such as the expected network traffic, the desired level of QoS that the network should provide, and so on, which are disregarded by the current approaches to the TC problem. In the following, we clarify this point with some examples.

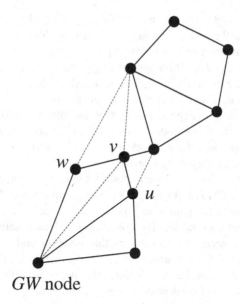

GW node

Figure 15.4 Example showing that a topology with optimal power spanning factor might not be optimal for extending network lifetime.

Consider the node placement reported in Figure 15.4, and assume we want to find an energy-efficient topology for this node placement. If we assume that the radiated power is the dominating factor in the node power consumption (this in general is not true – see Section 15.2.2 – but it is assumed that it holds here, for the sake of presentation), the power optimal topology is the Gabriel Graph (GG) (represented by solid edges in Figure 15.4), which has power spanning factor equal to 1. We recall that this property means that the energy-optimal path between any two nodes in the maxpower communication graph is always included in the GG, that is, no energy penalty is paid if communications take place only along the few links of the GG. Thus, from a graph-theoretic perspective, the GG is the energy-optimal topology.

Let us now consider the problem of identifying an energy-efficient topology under a more general perspective, that is, one of extending network lifetime (this is the ultimate goal of reducing node energy consumption). Is the GG still the best possible topology if we aim at extending network lifetime as much as possible? The answer to this question depends on the application scenario, which, in turn, determines which is the notion of lifetime we are interested in (e.g. time for the first node to die, or time to network disconnection, and so on), and also the expected traffic pattern in the network.

For instance, assume the network depicted in Figure 15.4 is a wireless sensor network used for environmental monitoring. In this scenario, sensor nodes periodically send data to one or more gateway nodes, which provide the interface to the user. Assume the gateway node is the one at the bottom left (see figure); then, the expected traffic pattern is one in which nodes periodically send data to this node, and a natural definition of network lifetime is 'time until no more meaningful data can be gathered from the network'.

Suppose nodes use only the links in the GG to send the data to the gateway node. As mentioned above, using the GG as the network topology allows nodes to send their data to the gateway along the most energy efficient path. Nevertheless, we claim that *the GG is not the optimal network topology if the goal is extending network lifetime.* In fact, if all the nodes send the data along the most energy efficient path, node v turns out to be a bottleneck since it relays all the data packets generated by all the upper nodes in the network. As a consequence of this, node v is expected to deplete its battery and die much faster than the other nodes in the network. When node v dies, most of the network nodes cannot send their data to the gateway node, and the network itself can be declared dead, as it is no longer able to provide the functionality it was designed for.

Let us now consider a relatively denser network topology, obtained by adding the dashed edges to the links in the GG (see Figure 15.4), and assume the routing protocol is energy aware, that is, it alternates the path selection in order to provide a balanced energy consumption among the network nodes. By combining a denser network topology with an energy-aware routing protocol, we can ensure that nodes u and w are also used to relay the data packets generated by the upper nodes in the network. This way, node v is no longer a bottleneck, and its lifetime is considerably increased with respect to the previous scenario. Since the traffic load generated by upper nodes is now shared among three nodes, the gateway node can gather data from the entire network for a longer time, and network lifetime is increased with respect to the previous case. Thus, a topology that is relatively denser than the GG is preferable if the designer goal is extending network lifetime as much as possible.

Consider now an example in which QoS is the main concern. For instance, assume a WSN scenario in which the fast notification of an abnormal detected event to the external user is vital. Examples of such scenarios are WSN used for intrusion detection, or to monitor nuclear plants. In this context, the network must fulfill a stringent QoS requirement, which can be expressed as a guarantee on the delivery time of the alarm message to the gateway node. What is the optimal network topology and/or routing strategy in this context?

Again, the answer to this question depends on features such as expected node density, expected traffic pattern, and so on. Let us consider two scenarios: in scenario (*a*), the network is very dense (many nodes per unit area) and the network traffic is moderate (for instance, because nodes exchange many packets to monitor the proper functioning of the plant); in scenario (*b*), the network is relatively sparse (few nodes per unit area) and the traffic is very low, except for occasional bursty periods (this is typically the case of a WSN used for intrusion detection). We claim that the optimal network topology/routing strategy is radically different in the two scenarios.

Let us first consider scenario (*a*). In this case, the best solution is probably to use an interference-optimal network topology, coupled with an interference-aware routing protocol.[2] In fact, assume a certain node u has detected an event that must be promptly notified to the gateway node v (see Figure 15.5). Node u can decide to either send the packet at maximum power, reaching node v through node w (two-hop path), or to send the packet at a lower power through the relatively longer path $\{s, t\}$, which turns out to be interference optimal. In principle, we might expect that sending the packets along the shortest path experiences the lowest delay (two hops instead of three hops are used to deliver the

---

[2]Interference-aware routing protocols send packets along low-interference paths in the network (Couto et al. 2003; Jain et al. 2003).

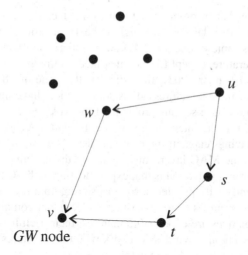

Figure 15.5 Example showing that finding the optimal network topology/routing strategy for minimizing packet delivery time depends on node density and expected network traffic.

packet); however, this does not account for interference with other nodes, which might cause transmission failures and repeated packet transmission. Since network traffic in this scenario is expected to be moderate, and node w resides in a densely populated region of the network, it is likely that node w experiences much higher interference than nodes s and t. So, sending the packet through the low-interference path $\{s, t\}$ is likely to ensure a shorter packet delivery time with respect to sending it through node w, despite the higher hop count of this path.

In scenario (b), the situation is opposite: although node w resides in a densely populated region of the network, the expected network traffic is very low, and, consequently, the interference at node w is likely to be very low (say, comparable to the expected interference at nodes s and t). In this situation, sending the packet through the shortest path is then the choice that most likely minimizes the packet delivery time.

The examples reported in this section have shown that the problem of identifying energy-optimal and/or interference-optimal network topologies is a very difficult one and that this problem can be properly tackled only by using a general approach that optimizes the network topology *and* the routing strategy as a function of certain target properties (e.g. extending network lifetime, or minimizing packet delivery time). We believe much work along this line of research is still to be done.

15.5 Implementation of TC

Despite the considerable body of research performed on topology control, and the many theoretical and simulation-based evidences of its potential of reducing node energy consumption and radio interference, to date there is no *experimental* evidence that TC techniques can be actually used to these purposes. The lack of experimental demonstrations of the usefulness of TC mechanisms is probably the most important open issue in this research field.

We want to outline that in order to implement TC techniques in a test bed, one must address the fundamental issue of how to integrate the TC functionality in the protocol stack. We have discussed this topic in Section 3.4, outlining that there is no clear proposal in this sense in the current literature, except for the integrated routing/power-control protocol introduced in (Kawadia and Kumar 2003). In particular, the issue of efficiently integrating TC with the MAC protocol must be addressed, as some authors have noticed that using asymmetric power levels may cause serious problems if the MAC layer is not properly designed (Jung and Vaidya 2002). If the interaction between TC and MAC is not carefully designed, the potential energy savings/capacity increases achieved by TC might be overweighed by undesired behaviors at the MAC layer, thus impairing the usefulness of TC mechanisms.

Another issue we want to mention is that experimenting with TC techniques on a test bed requires facing apparently silly problems, such as setting up a network topology composed of an adequate number of nodes (say, at least few tenths) that communicate along *multihop* paths. In fact, TC techniques are expected to manifest their usefulness in medium- to large-scale networks, and setting up a wireless network with these features might be very difficult in practice because of cost as well as logistic reasons.

Part VI

Case Study and Appendices

16

Case Study: TC and Cooperative Routing in Ad Hoc Networks

As outlined in Chapter 1, many problems are still to be solved before ad hoc and sensor networks can actually be used on a large scale. In this book, we have focused our attention on TC techniques, which can help the network designer to address two major difficulties: efficient use of the scarce energy resources and reduction of the radio interference level. In this chapter of this book, we consider another fundamental issue that arises in the context of ad hoc networks: how to stimulate cooperation between nodes in a competitive scenario. After briefly describing the issue of cooperation in ad hoc networks and introducing basic concepts from game theory, which will be used in the remainder of this chapter, we will show how the integration of the TC functionality with routing can lead to the design of simpler and more-efficient protocols (with respect to state-of-the-art solutions) to route messages in a competitive network.

16.1 Cooperation in Ad Hoc Networks

In a typical ad hoc network application scenario, network nodes are owned and managed by different authorities: private users, professionals, profit and no-profit organizations, and so on. This is the case, for instance, of ad hoc networks used for providing fast traffic information delivery in highways and metropolitan areas, or for providing ubiquitous Internet access (see Chapter 1 for a short description of these application scenarios).

When network nodes are owned by different authorities, a voluntary and 'unselfish' participation of the nodes in the execution of a certain networkwide task cannot be taken for granted. In fact, it is well known from economy that when humans are immersed in a competitive environment they tend to act selfishly in order to preserve their individual utility.

Selfish node behavior can have a dramatic impact on the performance of ad hoc network protocols, unless adequate countermeasures are taken at the design stage. We clarify this

Topology Control in Wireless Ad Hoc and Sensor Networks P. Santi
© 2005 John Wiley & Sons, Ltd

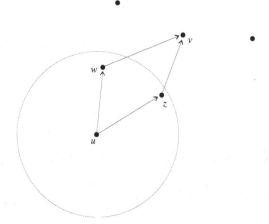

Figure 16.1   Example showing the disruptive effect of selfish node behavior on packet forwarding. Node u wants to send a packet to node v: since v is out of u's maximum transmitting range, only multihop communication through either node w or z is possible. However, in a competitive, selfish environment, w and z have no interest in forwarding u's packet, and the communication between u and v is not possible.

point with an example, which refers to the most basic service a multihop ad hoc network must provide, that is, routing of data packets.

Consider the situation depicted in Figure 16.1: the network is composed of a few nodes, and the only way for node u to send a packet to node v is to use either node w or node z as relay. If we consider a competitive environment and assume that node u is owned by authority A_1, node v is owned by authority A_2, and nodes w and z are owned by authority A_3, the communication between nodes u and v will never take place. In fact, nodes w and z have no interest in forwarding u's packet because, in general, only the sender and/or the receiver of a packet will get some utility from the communication. Indeed, nodes w and z have interest *in not* forwarding the packet since this action has only the effect of reducing their own resources (energy, and available bandwidth). Thus, either one of nodes w and z will simply drop any packet received from u, and the communication between u and v is not possible.

As the example above has shown, if some of the nodes in the network act selfishly, few multihop communications can take place, and the network functionality is compromised. Similar observations apply to other network services, such as information broadcast, P2P file sharing, multihop Internet access, and so on. In order to mitigate the disruptive effects of selfish node behavior, several authors have proposed techniques to stimulate cooperation between competing nodes in performing a certain networkwide task.

In the remainder of this chapter, we will focus our attention on the task of performing multihop message forwarding, which, given its vital role in the implementation of any ad hoc network service, is the most widely studied case in the literature. Recently, researchers have also considered the issue of cooperation in other networkwide tasks, such as multicast routing (Wang and Li 2004), wireless hot spots bandwidth sharing (Efstathiou and Polyzos 2005), and so on.

The most natural way of stimulating cooperation between selfish nodes in packet forwarding (and, indeed, in any other network service) is to use incentives, which can take the form of either *reputation systems* or *monetary transfer.*

In the former case, network nodes monitor each other's behavior; if some nodes are detected as not behaving according to the protocol specifications, they are marked as 'bad guys' and their identity is advertised to the other nodes so that they can be isolated from the rest of the network. In this approach, the incentive to behave according to the protocol specification (i.e., in a sense, to act unselfishly) is given by the fact that a deviation from the intended behavior (e.g. dropping someone else's packets), although possibly leading to a short-term utility increase, is likely to be detected by the other nodes, which exclude the bad-behaving node from the network. If this happens, the bad-behaving node's utility drops dramatically (for instance, because it can no longer send its own packets). So, a 'rational' node prefers to act unselfishly in order to avoid the high risk of being excluded from the network. Examples of reputation systems designed to stimulate cooperation in packet forwarding are (Buchegger and LeBoudec 2002a,b; Marti et al. 2000; Michiardi and Molva 2002).

In approaches based on monetary transfer, a node wishing to send a packet must pay a certain amount of money to the relay nodes to motivate them to forward its message. The sender of a packet accepts to pay some money for the service received (packet forwarding and delivery to the destination) since it expects to get some utility from the communication. In turn, the relay nodes are motivated to forward the packet since the money they receive as a compensation for their service can be used to pay the delivery of their own packets. Examples of routing protocols based on monetary transfer are (Anderegg and Eidenbenz 2003; BenSalem et al. 2003; Buttyan and Hubaux 2001, 2003; Eidenbenz et al. 2005; Zhong et al. 2003).

16.2 Reference Application Scenario

Although in principle the approach discussed in this case study can be used for establishing a generic connection between arbitrary source/destination node pairs in a competitive environment, in the following, we will specialize our techniques for a reference application scenario, which we describe briefly.

Assume a wireless network is used to access a certain service, such as a connection to the Internet through a wireless base station. A typical example is a network used to provide wireless Internet connection in coffee bars, airports, shopping centers, and so on.

In principle, multihop wireless communication (i.e. ad hoc networking) can be used to extend the area covered by the service, as depicted in Figure 16.2: instead of requiring each customer (also called *user*, or *node*, in the remainder of this chapter) to have direct connection with the base station, users can be allowed to reach the base station through multihop paths, using the wireless devices (laptops, PDAs, cell phones, and so on) of other customers as relay.

In order for this scenario to be successfully implemented, the nodes that are already accessing the service must be motivated to act unselfishly, cooperating to successfully realize

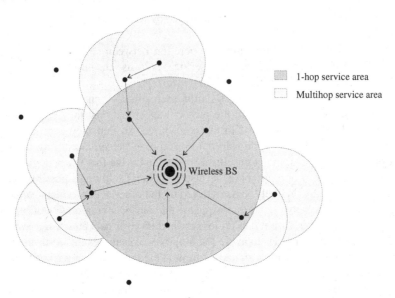

Figure 16.2 Multihop communication can be used to extend the coverage area of a certain wireless service (e.g. Internet access).

the task of connecting a new customer to the service. In fact, in this context, the connected nodes in principle have interest only in not allowing new nodes to join the service, as newcomers would subtract network resources (energy and available bandwidth) to the currently connected nodes.

To address this problem, a new node willing to access the service is requested to pay a certain amount of money to the nodes that will relay its packets. Clearly, the newcomer wants to pay the minimum amount of money needed to establish a successful connection to the access point, and, to this purpose, it launches a reverse auction[1] in which all the currently connected nodes participate. We will clarify this mechanism in the reminder of this chapter.

Note that in the scenario at hand both route discovery and packet forwarding must be implemented in order to connect a newcomer to the service provider. Thus, we need to device a mechanism that stimulates node cooperation in performing both these tasks.

For the sake of presentation, in the following, we conventionally call the customer who wants to establish a connection to the service the *sender*, the intermediate wireless nodes the *relays* (or *intermediate nodes*), and the service provider the *destination* of the communication, independent of the actual data flow between the sender and the destination. For instance, in case the accessed service is Internet connection, most of the traffic is likely to be downlink (i.e. from the base station to the customer). Nevertheless, the data session is initiated by the newcomer with a route (or service) discovery phase, and the newcomer will pay for both the incoming and outgoing traffic. For this reason, we adopt the terminology introduced above.

---

[1]The auction is reverse because the newcomer wants to buy, and not to sell, a service.

16.3   Modeling Routing as a Game

Most of the approaches proposed in the literature to stimulate cooperation between nodes, and the approach considered in this case study, have been designed and analyzed by borrowing techniques and concepts from a theory that is very well known in Computer Science and Economics, namely, *game theory* (see (Osborne and Rubinstein 1994) for an introduction to this subject).

Game theory is aimed at modeling and helping in understanding the phenomena that occur when several decision-makers (called *players*, or *agents*, in game theory terminology) interact. The assumptions at the basis of this theory are that players pursue well-defined objectives (they are *rational*), and take into account their knowledge or expectations of other players' behavior when performing their own decisions (they *act strategically*). Although these assumptions (especially the assumption that agents are rational) are sometimes criticized by economists, they allow a sound mathematical treatment of the subject, which would otherwise be very difficult to analyze.

In the remainder of this chapter, we model the process of *establishing a route* and *forwarding data packets* between a source and a destination node in a competitive environment as a game.

The players are the network nodes, which, with respect to a given data session, have one of the following roles: *source*, *relay* (or *intermediate*) node, and *destination*. In the reference scenario considered in this chapter, we assume that the source and the destination node are unique (unicast communication); hence, in a network composed of n nodes, the remaining $n - 2$ nodes play as (potential) relay nodes. As mentioned previously, the case of multiple destinations (multicast communication) has been recently considered in the literature (Wang and Li 2004).

The sender S has a private information (its *type*), which models its willingness to pay for establishing a connection to the service provider. By assuming that the sender can quantify this willingness to pay in monetary terms (this is a standard assumption in game theory), the sender's type corresponds to a maximum per-packet price m, which S will accept to pay. The *utility* u_S of the sender can then be modeled as

$$u_S = m - c_S(D),$$

where $c_S(D)$ is the actual per-packet amount of money that S will pay to establish a communication with D. In case the connection cannot be established (because of lack of wireless connection to the base station, or because of excessive cost for establishing the connection), the sender will get 0 utility.

Let us now consider an arbitrary relay node v. In this case, the private type of the player is its cost (which is also expressed in monetary terms) for forwarding S's packet. In general, the cost c_v incurred by node v for forwarding the packet depends on several factors, such as battery level, transmit power level (which, in turn, depends on the neighbor to which the packet is destined), bandwidth currently used by the node for its own connections, and so on. However, the only thing that is relevant here is that v's willingness to relay S's packet can be expressed as an amount of money c_v, independent of how c_v is computed. The utility of the generic relay node v can then be modeled as

$$u_v = Pay(v) - c_v,$$

where $Pay(v)$ is the per-packet payment that node v receives for relaying S's packet. In case node v does not take part in the data session, it receives no payment and incurs no cost, that is, it gets 0 utility.

Let us finally consider the third type of player in the game, the destination D. In the reference scenario described in Section 16.2, it is reasonable to assume that the service provider is a trustworthy third party, which has no interest in cheating as this would compromise the level of customers' confidence on the service. So, in our model, the destination node is not actually part of the game, but it is rather a 'neutral referee' whose goal is to correctly compute the cost-optimal (S, D) path, the payment for S, and the premiums for the relay nodes.

The assumption that one of the agents in the game is actually a trustworthy third party is quite common in the literature on stimulating cooperation in ad hoc networks, and it is also commonly used in the game theory literature. For instance, when analyzing the properties of an auction protocol (which is usually done by using game theoretic approaches), it is assumed that the auctioneer acts honestly when determining the winners of the auction and the price they must pay (Mas-Colell et al. 1995).

Given the assumption of rational behavior, the goal of each player (excluding the destination) is to maximize its own utility function. To this end, each player chooses to play a certain strategy that, in the reference scenario of Section 16.2, might include declaring a false type, dropping/modifying messages, and so on. One of these possible strategies is to follow the protocol specifications that, as we will see shortly, prescribe that players declare their true types, and to send/relay control and data messages. Using game theory terminology, we call this strategy *truth telling*.[2]

On the other hand, the goal of the protocol designer is to devise a mechanism such that a globally desirable goal is achieved (called the *social choice function* in game theory), and the individual utility of each node is maximized. Indeed, this is a sort of reverse engineering problem, which is known as *mechanism design* in the game theory literature (see Chapter 23 in (Mas-Colell et al. 1995) for an introduction to economic mechanism design, and the seminal paper (Nisan and Ronen 2001) on algorithmic mechanism design).

In a typical mechanism design problem, the goal is to define the game rules (including the payments) in such a way that the global outcome of the game played by the independent, utility maximizing agents corresponds to the desired social choice function. This is typically achieved by defining the game rules in such a way that truth telling becomes the *dominant strategy* for each player. A dominant strategy is one that maximizes the utility of the player no matter what strategies the other agents play. A mechanism such that truth telling is the dominant strategy for each player is called *truthful*, or *incentive compatible*, or *strategy proof*.

We remark that truthfulness is a very strong property, since it ensures that even if a player P has complete knowledge of other players' types and strategies truth telling is still the strategy that optimizes P's utility function. So, player P (and any other player in the game) has no incentive in deviating from the protocol specifications since deviating from the specifications would not lead to any utility increase. In the next section, we will give a practical interpretation of truthfulness in the reference scenario of Section 16.2.

---

[2]The reason for this name is the following. In standard, nondistributed game theory, a player's strategy often simply consists in declaring her type. Then, the strategy in which the player behaves honestly is called *truth telling*. By analogy, we call truth telling as the honest player behavior in the context of distributed game theory also.

Another important property the protocol must satisfy is *individual rationality*, which can be informally described as follows: a node is always better off participating in the game (i.e. in the protocol execution). Speaking in terms of utility function, individual rationality is satisfied if a node that executes the protocol never gets a negative utility. Since a player who does not take part in the game gets 0 utility, individual rationality ensures that participating nodes are not exposed to the risk of decreasing their utility. As a consequence, a rational player is motivated to join the game.

The only thing left to define in our routing game is the social choice function. The most natural option is to define the social optimum as 'establish the communication between S and D along the most efficient path', that is, along the path of minimum cost. In fact, communicating along the most efficient path on one hand optimizes network resources and, on the other hand, maximizes the likelihood of actually establishing a connection between the newcomer and the access point. With respect to this last point, we observe that, in order for a connection to be established, the sender has to pay an amount of money that is at least sufficient to cover the costs incurred by the forwarding nodes (this is necessary to motivate the relay nodes to forward the packet). Thus, the lower the overall path cost, the lesser the sender is expected to pay, the most likely it is that the sender actually accepts to establish the connection (we recall that the sender wants to spend at most m for sending a packet).

Summarizing, our goal is to design a truthful and individually rational protocol for route discovery and packet forwarding in ad hoc networks, such that the source and destination communicate using the path of minimum cost. Another desirable goal, which is not related to the game theoretic setting, is to reduce the protocol message overhead as much as possible.

To conclude this section, we want to mention that in our model we do not consider *collusion between nodes* and *malicious node behavior*. With collusion, we mean that a group of nodes (called a *coalition* in game theory terminology) can coordinate their cheating behavior, if this results in an increase of the overall coalition utility. The surplus obtained with this coordinate action can be shared among the nodes in the coalition, which then have an incentive to deviate from the truthful behavior. In case of malicious nodes, agents are allowed to play any strategy (also an irrational one), as long as this is detrimental to the system. Typically, the techniques used to avoid collusion between nodes and to prevent malicious node behavior are very different from the ones used to address noncollusive selfish node behavior. For this reason, in the following, we will focus our attention on noncollusive selfish nodes only.

16.4   A Practical Interpretation of Truthfulness

Before proceeding to describe the routing protocol, we want to explain why designing a truthful mechanism is so important in this context. To this purpose, we briefly discuss how truthfulness can be interpreted in a practical sense in the application scenario of Section 16.2.

When a new customer subscribes to the service, the provider will give her a hw/sw 'access kit', which implements at least the routing protocol used for establishing the connection. As part of the subscription, the new customer will also receive some information on how the service works: there is a fixed (monthly and/or per connection) fee, plus a variable per-connection cost. However, a customer can also be paid when other customers use her device for their connections.

With this type of service, a selfish user might be tempted to manipulate the 'access kit' in order to increase the payments she receives. However, if the mechanism is truthful, the user will never increase her utility by manipulating the 'access kit', since this kit is, in fact, an agent that is acting in the best possible selfish way on the customer's behalf. So, the user is not motivated to try to fool the system.

Another observation is that if the 'access kit' is truthful, it is in the interest of the service provider to give the customers very detailed information on the payment scheme: the amount of money a customer spends for establishing a connection and the premium received when a customer's device acts as a relay. Advertising this information is fundamental to convince even the expert user that cheating is not attractive.

16.5   Truthful Routing without TC

Before presenting a solution to the design problem described in Section 16.3, which integrates routing and topology control, in this section, we present the ADHOC-VCG routing protocol introduced in (Anderegg and Eidenbenz 2003), which is the only protocol introduced in the literature so far with the following features:

- it performs route discovery;

- it sends packets along the minimum-cost path;

- it is formally proven to be truthful and individually rational.

In fact, the other solutions to the problem of routing in selfish environments presented in the literature are concerned only with packet forwarding: in other words, it is assumed that the source/destination routes are somehow known to the nodes, and the problem is one of motivating nodes to forward packets. Furthermore, for some of these proposed protocols, no formal proof of resilience to strategic node behavior is presented. Since in our reference scenario performing both route discovery and packet forwarding is needed in order to establish a new connection, ADHOC-VCG is basically the only solution we can use.

ADHOC-VCG is based on the following idea: when a node (the sender S) wants to send a packet (to establish a connection with the access point in our reference scenario), it initiates a route discovery phase, declaring the destination of its packets (the access point in our case). At the end of the route discovery phase, the sender receives a message indicating the path P to the destination and the per-packet cost of sending (or receiving) a packet along P. The price p paid by the sender to send a packet is divided among the nodes in P, according to well-defined rules. Note that p must be at least sufficient to cover the cost of path P; otherwise, at least one node in the path would have no incentive for forwarding the packet. In the source payment model presented in (Anderegg and Eidenbenz 2003), S pays the premiums also (i.e. the amount of money exceeding the actual forwarding cost) to the relay nodes.[3]

---

[3]In the other payment model considered in (Anderegg and Eidenbenz 2003), the central-bank model, all the premiums due to network nodes are accumulated by a central authority, which divides them equally among all the nodes.

Although AdHoc-VCG has the nice features described above, it has also major shortcomings.

The first shortcoming is the high message overhead generated by the protocol execution, which is in the order of $O(n^3)$, where n is the number of network nodes. The high AdHoc-VCG overhead is due to the fact that this protocol is based on a link-weighted graph model. In other words, it is assumed that each link in the maxpower communication graph is weighted with the cost of performing the communication along it, and the minimum-cost path is calculated on the link-weighted graph. Since AdHoc-VCG is a fully distributed protocol, the cost of a link is calculated by probing each link individually, which incurs a considerable message overhead.

The second and more serious shortcoming is that AdHoc-VCG is proven to be truthful and individually rational under the assumption that both the destination and the sender act honestly. Differing from the model described in Section 16.3, it is then assumed that also the sender does not take part in the game: it behaves well by assumption, and it is not then required to satisfy individual rationality. We believe the assumption of honest sender's behavior is quite unrealistic in our reference scenario, as in this scenario many nodes are expected to act as both sender (when sending/receiving their own packets) and relay node (when forwarding other customers' packets) at the same time. Thus, the assumption of honest sender behavior implies that a node would behave strategically when forwarding someone else's packets, but it would become a 'good guy' (no strategic behavior) when it sends its own packets.

Note that the fact that the sender is not part of the game implies that S has no utility function to maximize. In fact, in AdHoc-VCG, it is implicitly assumed that once the sender has initiated the route discovery phase, it always pays the requested amount of money, that is, it cannot withdraw the communication request. Considering our reference scenario, the above assumption would imply that a customer, once issued the connection request, would be forced to pay an amount of money that she does not know in advance (and that could render its utility function negative). So, executing AdHoc-VCG is not individually rational for the sender: nobody would use a service whose cost is not known in advance!

16.6 Truthful Routing with TC

In this section, we present the COMMIT protocol for the truthful and cost-efficient routing introduced in (Eidenbenz et al. 2005). The novelty of this protocol with respect to AdHoc-VCG is that it is assumed that COMMIT is executed in combination with a TC protocol that assigns a specific power level to every node in the network.[4] Since a node v uses the same power level independent of the actual neighbor to which the packet is sent, we can assume that c_v does not depend on the packets next hop in the path. In other words, COMMIT uses a node-weighted graph model to compute the minimum-cost path, instead of a link-weighted graph model as used in AdHoc-VCG.

The use of a node-weighted instead of a link-weighted graph model is the cause of the better performance of COMMIT compared to AdHoc-VCG in terms of message overhead: in fact, COMMIT exchanges at most $O(n^2\Delta)$ messages to compute the optimal (S, D)

---

[4]Although in (Eidenbenz et al. 2005) it is not specified which TC protocols can or cannot used in combination with COMMIT, it turns out that any periodical, level-based TC protocol can be used for this purpose.

path, where Δ is the maximum node degree in the communication graph computed by the TC protocol. Considering that most TC protocols build communication graphs with small degree ($\Delta = O(\log n)$, or even $\Delta = O(1)$ in some cases), this is a notable improvement over the $O(n^3)$ messages exchanged by ADHOC-VCG.

COMMIT also solves the second shortcoming of ADHOC-VCG, since it is shown to be resilient to strategic sender behavior also. In other words, COMMIT is proven to be truthful in the scenario in which the sender is also part of the game, that is, it is a player whose goal is to maximize her own utility function. Furthermore, executing COMMIT is proven to be individually rational for the sender also. These two properties are achieved by using a different pricing scheme with respect to that used in ADHOC-VCG.

16.6.1   The COMMIT routing protocol

The COMMIT routing protocol is based on the following simple idea, which we explain in the context of our reference scenario. When a new customer wants to establish a connection to the service provider, she issues a 'connection request', stating the maximum amount of money she is willing to pay for it. The customer is committed (this explains the name of the protocol) to the connection request: if a connection to the access point with a cost less than (or equal to) the declared price can actually be established, the newcomer cannot withdraw the request, and she must pay the requested amount of money. However, if the price requested for the connection does exceed the declared price, the newcomer can withdraw the service request, incurring no cost. This way, the potential new customer has always full control of the maximum amount of money she will pay for the requested service, as it is reasonable to assume in a realistic scenario.

The implementation of the route discovery phase in COMMIT is very similar to the one performed in ADHOC-VCG, except for the fact that we do not need to compute weights on each link: when the sender wants to establish a connection to the destination, it sends a *route discovery* message, which contains the sender ID, the destination ID,[5] and the maximum amount m of money the sender is willing to pay for sending a packet to D. When an intermediate node v receives a route discovery message generated by S, it includes in the message its own ID and its cost c_v for forwarding the packet (we recall that in COMMIT it is assumed that the cost is associated with nodes, and not with links), and it propagates the route discovery message with the new information included. Note that both the sender and the intermediate nodes use the transmit power level set by the TC protocol to send packets, independent of the actual neighbors to which the packet is destined. If the communication topology is connected, one or more of the route discovery messages generated by the sender will eventually reach the destination node, which will compute the minimum-cost path MP, the payment for the sender, and the premiums for the nodes in MP. This information is sent back to the sender through the reverse MP path. After the route discovery phase is completed, the sender, provided the requested price does not exceed m, sends the packets along MP, including the payments for nodes in MP. For a detailed description of COMMIT implementation, the reader is referred to (Eidenbenz et al. 2005).

---

[5]In the reference application scenario at hand, including the destination ID in the message is not necessary as the destination is always the access point. However, in case the destination of a packet is arbitrary, including the destination ID in the route discovery message is necessary.

16.6.2   The COMMIT pricing scheme

Let us now describe the core of the COMMIT protocol, that is, the pricing scheme. We recall that we must devise a pricing scheme such that the following hold:

- *Truthfulness*: Cheating (e.g. declaring a false type) is not attractive for both the relay nodes and the sender.

- *Individual rationality*: Both the sender and the relay nodes participating in the protocol execution never get a negative utility.

- *Cost efficiency*: If feasible, the communication takes place along the path of minimum cost.

From game theory, it is known that any pricing scheme that satisfies the three properties above must be based on the VCG mechanism (Feigenbaum et al. 2002). The VCG mechanism, named after Vickrey (Vickrey 1961), Clarke (Clarke 1971), and Groves (Groves 1973), is also used in the ADHOC-VCG protocol, and defines the payments to the nodes as follows.

Definition 16.6.1 (VCG payments) *Let $c(P)$ denote the cost of an arbitrary (S, D) path P, where $c(P) = \sum_{v \in P, v \neq S, D} c_v$, and let MP denote the path of minimum cost connecting S and D in the communication graph. For any $v \neq S$, D in MP, let $c(P^{-v})$ denote the cost of the optimal (S, D) path P^{-v}, which does not include node v. Path P^{-v} is called the replacement path for node v. The payment for node v is defined as*

$$Pay(v) = c(P^{-v}) - c(MP) + c_v,$$

where c_v is the declared forwarding cost of node v. The payments for the nodes that are not on the minimum-cost path are set to 0.

Note that the above definition of VCG payments assumes a node-weighted graph model. By redefining the cost of a path as the sum of the link costs, we can extend the notion of VCG payments to the case of link-weighted graphs (this is the model used in the ADHOC-VCG protocol).

The intuition behind the VCG pricing scheme is the following: the payment a relay node v receives exceeds its forwarding cost by a positive quantity $c(P^{-v}) - c(MP)$, which measures the 'added value' provided by node v to the network: was node v not part of the network, the minimum cost for connecting S to D would be $c(P^{-v}) \geq c(MP)$. Thus, $c(P^{-v}) - c(MP)$ represents the added value provided by node v to the network, and the node is given exactly this amount of money in excess of its forwarding cost.

Note that the cost c_v used in the definition of the VCG payments is the cost *declared* by node v, which could be different from the actual forwarding cost c'_v of node v. However, the following informal argument shows that declaring a false type does not increase a node's utility with VCG payments (for a formal proof of this, see (Anderegg and Eidenbenz 2003)).

Suppose node v declares a forwarding cost $c_v = c'_v + \varepsilon$, and that v remains in the minimum-cost path MP also with this false declaration. Since the increase in the declared cost causes an equivalent increase in the cost of MP (this is true under the assumption that the other nodes in the path do not change their declarations), and $c(P^{-v})$ does not change (because the cost declared by node v has no influence on the cost of the optimal path not including v), the payment that node v will receive remains unchanged. In fact,

$$Pay(v) = c(P^{-v}) - (c(MP) + \varepsilon) + (c'_v + \varepsilon) = c(P^{-v}) - c(MP) + c'_v,$$

which is the payment that node v would have received by declaring its true type c'_v. A similar argument applies to the case in which node v underdeclares its true forwarding cost. Since by under- or overdeclaring its packet forwarding cost node v cannot increase its utility, but these actions only expose the node to the risk of being excluded from the minimum-cost path, it follows that v has no incentive in declaring a false type. In other words, VCG payments can be used to design a truthful routing protocol.

As for individual rationality, we observe that a node v that is part of the minimum-cost path will always receive an amount of money that is at least equal to its forwarding cost, that is, it has a positive utility. In case node v is not in the minimum-cost path, it gets no payment and incurs no cost, that is, it has 0 utility. Thus, participating in the game is individually rational for node v, since this action can only increase (or leave unchanged) node v's utility.

As the example reported in Figure 16.3 shows, with VCG payments, the sum of the premiums paid to the intermediate nodes, in general, exceeds the true cost of sending a packet from S to D. In the node configuration reported in Figure 16.3, the minimum-cost path connecting S and D is $MP = \{S, u, v, z, D\}$, which has cost equal to 46. The VCG payments for the nodes in $MP - \{S, D\}$ are as follows:

$$Pay(u) = c(P^{-u}) - c(MP) + c_u = 53 - 46 + 15 = 22,$$

$$Pay(v) = c(P^{-v}) - c(MP) + c_v = 48 - 46 + 19 = 21,$$

$$Pay(z) = c(P^{-z}) - c(MP) + c_z = 48 - 46 + 12 = 14,$$

which sums up to $57 > 46$.

The problem described above, which is known as *budget imbalance* in game theory, is intrinsic in the VCG mechanism and, under reasonable assumptions, in any mechanism that achieves truthfulness, individual rationality, and cost efficiency at the same time (Mas-Colell et al. 1995). The reason for this economic inefficiency is that, in order to motivate

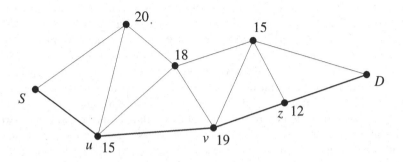

Figure 16.3   Example showing the budget imbalance problem with VCG payments. The minimum-cost path connecting S and D is $MP = \{S, u, v, z, D\}$ (bold edges), with cost equal to 46. The sum of the VCG payments to the nodes in $MP - \{S, D\}$ is $57 > 46$.

relay nodes to cooperate, they must be given premiums exceeding their forwarding cost. The difference between the overall amount of these premiums and the cost of the optimal (S, D) path has been called *the cost of cooperation* in the literature (Eidenbenz et al. 2005). Thus, the cost of cooperation in the example of Figure 16.3 is 11.

Differing from ADHOC-VCG, in COMMIT, we must also define the payment rule for the sender, in such a way that truthfulness and individual rationality are satisfied. The price $c_S(D)$ that S should pay for the communication defines the *decision rule*, which determines whether the communication takes place $(c_S(D) \leq m)$ or not $(c_S(D) > m)$.

A trivial choice would be to set $c_S(D) = \sum_{v \in MP, v \neq S, D} Pay(v)$, that is, to let the sender pay the overall amount of the premiums due to the nodes in $MP - \{S, D\}$. However, the following example shows that this definition of $c_S(D)$ leaves space for a strategic behavior of the nodes in the minimum-cost path.

Consider again the node configuration depicted in Figure 16.3, and assume the sender has a reserve price $m = 56$. If all the nodes in the network behave truthfully, the communication between S and D cannot be established, since the cost that S must pay is $c_S(D) = 57$, which exceeds the reserve price. Assume now that node z falsely declares a forwarding cost of 13, instead of 12. With this false declaration, the minimum-cost path is still $MP = \{S, u, v, z, D\}$, but the premiums paid to the nodes are changed as follows:

$$Pay(u) = c(P^{-u}) - c(MP) + c_u = 53 - 47 + 15 = 21,$$

$$Pay(v) = c(P^{-v}) - c(MP) + c_v = 48 - 47 + 19 = 20,$$

$$Pay(z) = c(P^{-z}) - c(MP) + c_z = 48 - 47 + 13 = 14,$$

which sums up to $55 < 56$, and the communication takes place. Thus, by falsely reporting its type, node z would be able to increase its utility from 0 (because the communication would not take place if z would report its true cost) to 2, which implies that the payment scheme obtained by setting $c_S(D) = \sum_{v \in MP, v \neq S, D} Pay(v)$ is not truthful.

To solve this problem, the authors of (Eidenbenz et al. 2005) suggest defining $c_S(D)$ as follows.

Definition 16.6.2 (Global replacement path and sender payment) *Given the minimum-cost path MP, the global replacement path* P^{-MP} *is the path of minimum cost that connects S and D in the communication graph and does not include any node in $MP - \{S, D\}$. Denoting by $c(P^{-MP})$ the cost of the global replacement path, the payment for the sender is set as* $c_S(D) = c(P^{-MP})$.

The defining of the sender payment as the cost of the global replacement path prevents the subtle strategic node behavior described above. In fact, with this definition of $c_S(D)$, the price paid by the sender (which determines whether the communication will actually take place) is no longer influenced by the declarations of the nodes in the minimum-cost path, and the forwarding nodes have no way of driving $c_S(D)$ below m, as in the example reported above. Referring back to the node configuration of Figure 16.3, the sender payment is $c_S(D) = c(P^{-MP}) = 53 < 56$, and the connection between S and D can be established.

While ensuring truthfulness and individual rationality for both the sender and the forwarding nodes, the sender payment rule as defined in Definition 16.6.2 introduces an additional budget balancing problem. In fact, when the connection is established, the sender

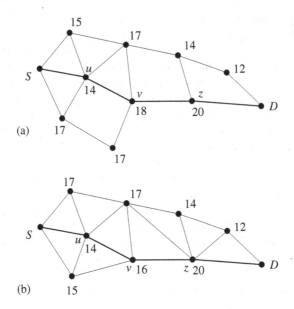

Figure 16.4 Examples showing that the cost of the global replacement path can be either lower (a) or higher (b) than the sum of the premiums due to the forwarding nodes. The minimum-cost path is represented by bold edges.

pays an amount of money that, in general, differs from the overall amount of the premiums due to the forwarding nodes.

As the examples reported in Figure 16.4 show, the cost of the global replacement path might be lower or higher than the sum of the premiums due to the forwarding nodes. In the node configuration at the top of the figure, the minimum-cost path is $MP = \{S, u, v, z, D\}$, and the payments to the forwarding nodes are $Pay(u) = 20$, $Pay(v) = 23$, and $Pay(z) = 25$, summing up to 68. However, the sender must pay only 58, which corresponds to the cost of the global replacement path. In the node configuration at the bottom of the figure, the minimum-cost path is again $MP = \{S, u, v, z, D\}$. In this case, the payments to the forwarding nodes are $Pay(u) = 15$, $Pay(v) = 17$, and $Pay(z) = 27$, which sum up to 59. The cost of the global replacement path is $c(P^{-MP}) = 60$, which, contrary to the previous case, exceeds the sum of the premiums due to the forwarding nodes.

To address this budget balancing problem, the authors of (Eidenbenz et al. 2005) suggest that the destination of the communication (the service provider in our reference scenario) should balance the budget, providing the additional money needed to pay the relay nodes in case the price paid by the sender is not sufficient to cover the forwarding expenses, or getting some money from the sender in case the price paid by S exceeds the forwarding expenses.

The payment model proposed in (Eidenbenz et al. 2005) seems quite reasonable in the reference scenario described in Section 16.2, since the service provider is likely to be involved in many communication sessions, and it is possible that its overall balance is close to 0. Even if this is not the case (e.g. because $c(P^{-MP}) < \sum_{v \in MP, v \neq S, D} Pay(v)$ on the average), the provider can act on the fixed price requested to subscribe to the service in order to preserve its revenue.

In application scenarios in which the destination is an arbitrary network node, the argument in favor of using the destination node to balance the budget is quite weak, since a node might not be willing to accept data from the sender if this requires paying some money. So, in these scenarios, we could use a different payment model, such as the central-bank model used in (Anderegg and Eidenbenz 2003), where a central authority balances the budget, equally dividing the cost (or the profit) of balancing the budget among the network nodes.

16.6.3 Protocol analysis

The following properties of the COMMIT protocol have been proven in (Eidenbenz et al. 2005):

- *Cost efficiency*: If all nodes act truthfully, COMMIT computes the minimum-cost route from S to D in the communication graph.

- *Message complexity*: The total number of messages exchanged by nodes executing COMMIT to establish a S, D route is $O(n^2 \Delta)$, where Δ is the maximum node degree in the communication graph.

- *Truthfulness*: Behaving truthfully is a dominant strategy for both the sender and the relay nodes.

- *Individual rationality*: Executing COMMIT is individually rational for both the sender and the relay nodes.

While a formal proof that the truthfulness property holds for the relay nodes is quite complicated (the interested reader is referred to (Eidenbenz et al. 2005)), the proof that behaving honestly is a dominant strategy for the sender is quite straightforward, and it is reported in the following.

We recall that, according to COMMIT specifications, the sender must: (i) declare its reserve price m; (ii) send a route discovery message that includes m; and (iii) in case the communication can be established at a price that does not exceed m, send the data along the computed path, and pay the requested connection price.

The sender might cheat in several different ways, such as,

1. declare a false reserve price m_f;

2. not sending the route discovery message;

3. not sending the data on the computed route;

4. pay less money than the requested price;

5. any combination of behaviors (1), (2), (3) and (4) above.

Let us first consider cheating option (1). Let m_f and m denote the false (the one included in the route discovery message) and the true type of node S, and assume $m_f < m$. The following cases can occur:

- $c(P^{-MP}) < m_f < m$: In this situation, the communication takes place with both declarations, and the utility of node S with the false declaration remains unchanged. This is because S's utility depends on $c(P^{-MP})$ (which is not influenced by S's declaration) and on the *true* type m.

- $m_f \leq c(P^{-MP}) \leq m$: In this case, if the sender would declare m_f instead of m, it would get 0 utility because the connection to the destination is not established. On the other hand, by reporting its true type, the sender would establish a connection to the destination, getting a positive utility of $m - c(P^{-MP})$. Then, declaring a false type decreases the utility of the sender.

- $m_f < m < c(P^{-MP})$: In this situation, the sender's utility is 0 with both declarations.

The above case-by-case analysis proves that underdeclaring the type is not attractive for the sender. By applying similar arguments, it can be easily shown that overdeclaring the type is also not attractive for the sender, which implies that cheating option (1) cannot increase the sender's utility.

Consider now cheating option (2). If the sender does not send the routing discovery message, it definitely cannot establish a connection with the destination, which results in 0 utility for the sender. Since by sending the route discovery message node S will get positive utility (we recall that the payment scheme ensures that S will never pay an amount of money that exceeds m), it follows that cheating option (2) is also not attractive for the sender.

As for cheating option (3), we observe that the sender can: (i) send the packet along a route different from the computed one, or (ii) not send the data at all. Situation (i) cannot occur, since the sender is aware of only one route to the destination, the one computed by COMMIT (which is the cost-optimal route). As for case (ii), we observe that the sender is expected to get some positive utility only if the data is actually delivered to the destination. So, not sending the data results in 0 utility for the sender, and also, this action cannot increase S's utility with respect to the case of honest node behavior.

In principle, cheating option (4) might increase node S's utility: by paying less money than the requested price $c_S(D)$, the sender apparently increases its utility. However, we remark again that the sender gets some positive utility *only if the data is actually delivered to the destination*. In the following, we prove that if the sender pays less than the requested amount of money, its packets will not reach the destination, implying that node S gets 0 utility from cheating option (4).

We have to consider the following cases:

- *The requested price $c_S(D)$ is less than or equal to the sum of the payments due to the forwarding nodes*: In this case, the destination node provides some money m' to balance the budget and pay all the forwarding nodes. However, if S provides less money than requested (say $m'' < c_S(D)$), we have that $m' + m''$ is less than the sum of the payments due to the intermediate nodes. So, one of these nodes will not receive the expected amount of money,[6] and, as a consequence, will drop S's packets. Thus, node S's utility in this situation is 0.

---

[6]Each node in the minimum-cost path is informed of the amount of its due payment when the computed route is sent back from the destination to the sender.

– *The requested price $c_S(D)$ exceeds the sum of the payments due to the forwarding nodes*: In this case, the lesser money m'' provided by the sender might be sufficient to cover the forwarding costs, so its packets can reach the destination. However, the destination node knows the amount of money in excess of the forwarding costs that it must get from the sender and, on receiving a lesser amount of money, it can decide to refuse S's packets. Thus, in this situation also, node S gets 0 utility.

Finally, by a straightforward case-by-case analysis, we can prove that any combination of cheating options (1), (2), (3) and (4) cannot increase S's utility. This proves that acting truthfully is a dominant strategy for the sender.

16.6.4 Interplay between TC and COMMIT routing

In this section, we discuss the interplay between TC and cooperative routing in ad hoc networks. As we shall see, on one hand implementing a truthful, individually rational, and cost-efficient routing protocol imposes some constraints on the underlying network topology. On the other hand, by carefully choosing the network topology, the designer can mitigate the budget imbalance problems induced by the requirement of implementing a truthful routing mechanism.

As we have observed in Section 16.6.2, it is known from game theory that any pricing scheme that satisfies truthfulness, individual rationality, and cost efficiency must be based on VCG-like payments. To determine the VCG payment of a node v on the minimum-cost path, we must compute the cost of the replacement path for v, that is, of the path of minimum cost connecting S and D in the communication graph that does not include node v. Of course, a prerequisite for computing this cost is that *the replacement path actually exists*. In other words, in order for the VCG payments to be correctly computed, the communication graph must satisfy the following property: for any node v in the graph, and for any source/destination pair (S, D), removing v from the graph does not disconnect nodes S and D. Using graph-theory terminology, this is equivalent to requiring that the communication graph is biconnected.

Indeed, the sender payment rule used in COMMIT, which is also based on the VCG mechanism, poses a stronger constraint on the communication graph. In fact, the definition of the sender payment requires computing the cost of the global replacement path which, we recall, is the path of minimum cost connecting S and D in the communication graph that does not include any of the nodes in $MP - \{S, D\}$, where MP is the minimum-cost path between S and D. Thus, in order to correctly compute the sender payment, the communication graph must satisfy the following property, which we call *minimum-cost biconnectivity*: for any source/destination pair (S, D), there always exists a path connecting S and D in the communication graph that does not include any of the nodes in $MP - \{S, D\}$, where MP is the minimum-cost (S, D)-path. Note that this graph property is stronger than biconnectivity. In fact, minimum-cost biconnectivity requires that one of the at least two node-disjoint paths that always exist between S and D (because of biconnectivity) is the minimum-cost path.

The difference between biconnectivity and minimum-cost biconnectivity is shown in Figure 16.5. The graph reported at the top of the figure satisfies biconnectivity (from the viewpoint of nodes S and D), but it is not minimum-cost biconnected: in fact, S and D become disconnected when all the nodes in $MP - \{S, D\}$ are removed from the communication graph. On the contrary, the graph reported at the bottom of Figure 16.5 satisfies

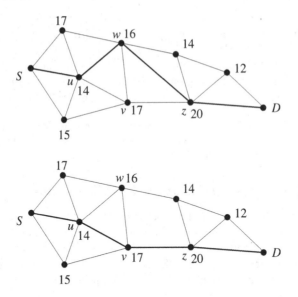

Figure 16.5 Examples showing the difference between biconnectivity and minimum-cost biconnectivity. The graph at the top is biconnected (from the view point of nodes S and D), but it is not minimum-cost biconnected: removing the nodes in the minimum-cost path (bold edges) disconnects S and D. The graph at the bottom satisfies minimum-cost biconnectivity: the minimum-cost path is $MP = \{S, u, v, z, D\}$ (bold edges), and removing nodes u, v and z leaves S and D connected.

minimum-cost biconnectivity, since removing all the nodes in $MP - \{S, D\}$ does not disconnect nodes S and D.

The above discussion indicates that the topology generated by a TC protocol to be used in combination with COMMIT routing must fulfill stringent requirements, that is, satisfying minimum-cost biconnectivity. Although the problem of building a k-connected topology (for some constant $k > 1$) has been recently studied in the literature (see (Bettstetter 2002; Li and Hou 2004; Penrose 1999a; Wan and Yi 2004)), the problem of designing a protocol that builds a minimum-cost biconnected topology has not been adequately addressed yet. The simulation results presented in (Eidenbenz et al. 2005) show that popular TC protocols such as CBTC (see Section 11.1) and KNEIGH (see Section 12.2) produce topologies that, with high probability, satisfy minimum-cost biconnectivity. The same result holds for the topology resulting when all the nodes have the same transmitting range, which equals the critical value for connectivity. However, these are average-case results based on simulations, and the problem of designing a TC protocol that builds a provably minimum-cost biconnected topology remains open.

Let us now consider the second issue we want to discuss, that is, mitigating the budget imbalance problem. As observed in Section 16.6.2, the COMMIT payment scheme induces two budget imbalance problems: one, which is known in the literature as the cost of cooperation, is due to the fact that the sum of the premiums paid to the intermediate nodes exceeds the actual packet forwarding cost; the second problem, which is induced by the sender payment rule, consists of the fact that the price paid by the sender in general is different from the sum of the premiums due to the forwarding nodes.

In general, the interest of the network designer is to reduce the budget imbalances described above as much as possible, as an imbalanced budget always results is some form of economic inefficiency. We explain this point with two examples.

Consider the reference application scenario of Section 16.2, and assume the service provider wants to design a truthful, individually rational, and cost-efficient routing protocol for accessing its service. The provider knows that, in order to motivate relay nodes to cooperate, they must be given a premium that exceeds their actual packet forwarding cost. However, it is in the interest of the service provider to reduce the extra money given to the intermediate nodes as much as possible. In fact, this extra money will be paid by either the sender or by the service provider itself (according to the payment model used in COMMIT). If it is paid by the sender, a lesser amount of extra money due to the relay nodes increases the likelihood of actually establishing a connection with the new customer (who, we recall, has a reserve price m for accessing the service); if the extra money is paid by the provider, its interests in reducing the cost of cooperation as much as possible is even more evident. One might observe that the service provider can balance the costs caused by a relatively high cost of cooperation by acting on the fixed fee requested of the nodes to access the service; however, increasing the fixed fee would generate no additional profit for the service provider (because increasing the fee is only a form of compensation for the high cooperation cost), while it potentially reduces the number of customers willing to access the service. On the other hand, relay nodes are happy when receiving even a very small amount of extra money, as this extra money anyway increases their utility with respect to not forwarding the packet. So, a reduced cost of cooperation does not compromise the routing mechanism, and improves the economic efficiency of the system.

Let us now consider the second budget imbalance problem, caused by the fact that the sender pays the cost of the global replacement path, which is, in general, different from the sum of the payments due to the forwarding nodes. As discussed in Section 16.6.2, in the COMMIT payment model, the service provider is assumed to balance the budget, either by receiving some money from the sender or by giving some money to the intermediate nodes. We claim that it is in the interest of the service provider that the price requested of (and possibly paid by) the sender is as close as possible to the actual packet forwarding cost (including the extra money to the relay nodes). In fact, if the price requested to the sender is excessively high, the service provider potentially receives a lot of money from the connected customers. However, it is most likely that many potential customers will refuse the connection because of the high cost, which results in a loss of money for the service provider. On the other hand, if the price requested to the sender is excessively low, the service provider incurs a considerable cost to balance the budget. As observed above, this cost can be somehow be compensated by increasing the fixed fee requested to the customers, but the fixed fee increase is likely to reduce the number of customers accessing the service, while leaving the provider's profit unchanged.

When the design of a cooperative routing protocol is addressed as a stand-alone task, there is little the designer can do to mitigate the budget imbalance problems described above. In fact, these problems are caused by VCG-like payment schemes, which, unfortunately, are basically the only possible option to achieve truthfulness, individual rationality, and cost efficiency at the same time. However, if the problem of designing a cooperative routing

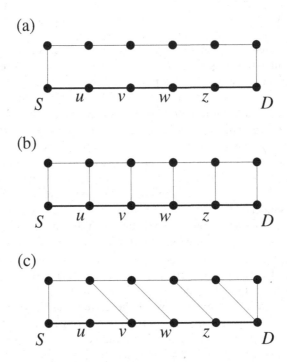

Figure 16.6 Examples showing how the careful choice of the network topology can consid-erably mitigate the budget imbalance problems. All the nodes have the same weight equal to 1, which is not reported in the graphs. The minimum-cost path is represented by the bold edges.

protocol is tackled using a wider perspective, which comprises both routing and TC, the network designer does have a handle to reduce budget imbalances: the careful design of the network topology on which the source/destination routes are computed.

We illustrate this point with an example. Consider the node configuration reported in Figure 16.6(a). For simplicity, assume all the nodes have the same forwarding cost equal to 1. In this situation, the minimum-cost path is the path of minimum hop count connecting S with D, that is, the path $MP = \{S, u, v, w, z, D\}$ with cost equal to 4. The payments to the nodes in $MP - \{S, D\}$ are as follows:

$$Pay(u) = Pay(v) = Pay(w) = Pay(z) = 6 - 4 + 1 = 3.$$

Thus, the total of the premiums paid to the forwarding nodes is 12, and the cost of cooperation is $12 - 4 = 8$. As for the second type of budget imbalance, we observe that the cost of the global replacement path is 6, which implies that the destination node must provide the remaining 6 units of money to balance the budget.

Let us now consider the node configuration reported in Figure 16.6(b). The network topology is now relatively denser as four new links are added to the communication graph. Intuitively, adding links to the communication graph should reduce both the cost of cooper-ation (because the 'added value' of a node in the minimum-cost path is expected to decrease when new links are added) and the amount of money provided/received by the destination node (because the difference between the cost of the global replacement path and the overall forwarding cost should be relatively smaller for relatively denser graphs). However, this happens only if the additional links are carefully chosen: in the node configuration reported in Figure 16.6(b), the minimum-cost path, the payments to the nodes, and the cost of the global replacement path remain unchanged, which implies that the cost of cooperation is still 8, and the destination node still has to provide 6 units of money to balance the budget.

Finally, consider the node configuration of Figure 16.6(c). Similar to the topology of Figure 16.6(b), this topology has four more links than the one reported in Figure 16.6(a). However, these four links turn out to be very useful to mitigate both types of budget imbalance, which are halved with respect to the case of Figure 16.6(a). In fact, the cost of the optimal path remains unchanged, but the payments due to the nodes in $MP - \{S, D\}$ change as follows:

$$Pay(u) = Pay(v) = Pay(w) = Pay(z) = 5 - 4 + 1 = 2.$$

As a consequence, the cost of cooperation is halved to $8 - 4 = 4$. Furthermore, the cost of the global replacement path with this node configuration is 5, which implies that the destination node must provide only $8 - 5 = 3$ units of money to balance the budget.

16.7 Conclusion

In this chapter, we have studied the problem of stimulating node cooperation in the fundamental task of message routing, and we have demonstrated the advantage of using a cross-layer approach to solving this problem: by exploiting the interplay between routing and TC, we have been able to design a cooperative routing protocol that generates less message overhead with respect to state-of-the-art approaches and successfully deals with strategic sender behavior. On the other hand, the investigation of the cooperative routing problem has disclosed new research directions in the field of TC: designing protocols that build a provably minimum-cost biconnected topology (possibly in a fully distributed and localized way), and characterizing 'good topologies' for the purpose of mitigating the inevitable budget imbalance problems caused by the use of a cooperative routing mechanism.

With respect to this last point, we observe that intuitively a 'good topology' for the purpose of cooperative routing should contain several paths of approximately the same cost between the source/destination pairs. In fact, if this happens, the expected 'added value' of the nodes in the minimum-cost path is relatively low, as is the expected difference between the cost of the global replacement path and the sum of the premiums due to the relay nodes. In order to have several paths of approximately the same cost, the network topology should be relatively dense, as the examples reported in Figure 16.6 show. This intuition is confirmed also by the simulation results reported in (Eidenbenz et al. 2005), which show that a relatively dense topology such as the one obtained when all the nodes have a transmitting range equal to the critical value for connectivity can reduce the magnitude of the budget imbalance problems by about 15% with respect to the case of relatively sparser topologies such as the ones generated by CBTC and KNEIGH. So, the goal of building a 'good topology' for the purpose of cooperative routing appears to be at least partially conflicting with the goals of reducing node energy consumption and radio interference. Further investigation along this research direction is needed.

Finally, we observe that in this chapter we have implicitly assumed that the nodes execute the TC protocol truthfully. This assumption is in general quite unrealistic, since a selfish node is expected to manipulate the TC protocol also, if this results in an increase of its utility. Thus, designing truthful TC protocols is also a very interesting open research topic, which has been only partially addressed by a recent paper (Eidenbenz et al. 2003).

A

Elements of Graph Theory

In this Appendix, we report basic definitions and concepts from graph theory that have been used in this book. Most of the material presented in this Appendix is based on (Bollobás 1998) (Section A.1) and on (Goodman and O'Rourke 1997) and (deBerg et al. 1997) (Section A.2).

A.1 Basic Definitions

Definition A.1.1 (Graph) *A graph G is an ordered pair of disjoint sets (N, E), where $E \subseteq N \times N$. Set N is called the vertex, or node, set, while set E is the edge set of graph G. Typically, it is assumed that self-loops (i.e. edges of the form (u, u), for some $u \in N$) are not contained in a graph.*

Definition A.1.2 (Directed and undirected graph) *A graph $G = (N, E)$ is directed if the edge set is composed of ordered node pairs. A graph is undirected if the edge set is composed of unordered node pairs.*

Examples of directed and undirected graphs are reported in Figure A.1. Unless otherwise stated, in the following by *graph* we mean *undirected graph*.

Definition A.1.3 (Neighbor nodes) *Given a graph $G = (N, E)$, two nodes $u, v \in N$ are said to be neighbors, or adjacent nodes, if $(u, v) \in E$. If G is directed, we distinguish between incoming neighbors of u (those nodes $v \in N$ such that $(v, u) \in E$) and outgoing neighbors of u (those nodes $v \in N$ such that $(u, v) \in E$).*

Definition A.1.4 (Node degree) *Given a graph $G = (N, E)$, the degree of a node $u \in N$ is the number of its neighbors in the graph. Formally,*

$$deg(u) = |\{v \in N : (u, v) \in E\}|.$$

If G is directed, we distinguish between in-degree (number of incoming neighbors) and out-degree (number of outgoing neighbors) of a node.

Figure A.1 Examples of directed graph (a) and undirected graph (b).

Definition A.1.5 (Path) *Given a graph $G = (N, E)$, and given any two nodes $u, v \in N$, a path connecting u and v in G is a sequence of nodes $\{u = u_0, u_1, \ldots, u_{k-1}, u_k = v\}$ such that for any $i = 0, \ldots, k - 1$, $(u_i, u_{i+1}) \in E$. The length of the path is the number of edges in the path.*

Definition A.1.6 (Cycle) *A cycle is a path $C = \{u_0, \ldots, u_k\}$ such that $k \geq 3$, $u_0 = u_k$, and the other nodes in C are distinct from each other and from u_0.*

Definition A.1.7 (Node distance) *Given a graph $G = (N, E)$ and any two nodes $u, v \in N$, their distance $dist(u, v)$ is the minimal length of a path connecting them. If there is no path connecting u and v in G, then $dist(u, v) = \infty$.*

Definition A.1.8 (Graph diameter) *The diameter of graph $G = (N, E)$ is the maximum possible distance between any two nodes in G. Formally,*

$$diam(G) = \max_{u,v \in N} dist(u, v).$$

Definition A.1.9 (Subgraph) *Given a graph $G = (N, E)$, a subgraph of G is any graph $G' = (N', E')$ such that $N' \subseteq N$ and $E' \subseteq E$. Given any subset N' of the nodes in G, the subgraph of G induced by N' is defined as $G_{N'} = (N', E(N'))$, where $E(N') = \{(u, v) \in E : u, v \in N'\}$, that is, $G_{N'}$ contains all the edges of G such that both endpoints of the edge are in N'.*

Definition A.1.10 (Symmetric sub- and supergraph) *Let $G = (N, E)$ be a directed graph. The symmetric subgraph of G, denoted G^-, is the graph obtained from G by removing all edges such that $(u, v) \in E$ and $(v, u) \notin E$. Formally, $G^- = (N, E^-)$, where $(u, v) \in E^-$ if and only if $(u, v) \in E$ and $(v, u) \in E$. The symmetric supergraph of G, denoted as G^+, is the graph obtained from G by adding the reverse edge to all unidirectional edges in G. Formally, $G^+ = (N, E^+)$, where $(u, v) \in E^+$ if and only if $(u, v) \in E$ or $(v, u) \in E$.*

Definition A.1.11 (Order of a graph) *The order of graph $G = (N, E)$ is the number of nodes in G, that is, the cardinality of set N.*

Definition A.1.12 (Complete graph) *The complete graph $K_n = (N, E)$ of order n is such that $|N| = n$, and $(u, v) \in E$ for any two distinct nodes $u, v \in N$.*

Definition A.1.13 (Sparse graph) *A graph $G = (N, E)$ of order n is sparse if $|E| = O(n)$, that is, if the number of edges in G is linear in n. If a graph is sparse, the average node degree is $O(1)$.*

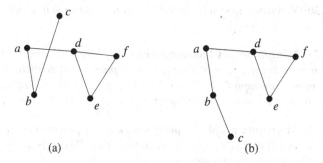

Figure A.2   Notion of graph planarity. The drawing of the graph $G = (\{a, b, c, d, e, f\}, \{(a, b), (b, c), (a, d), (d, e), (d, f), (e, f)\})$ in (a) is not planar; yet, graph G is planar, as shown by the drawing in (b).

Definition A.1.14 (Planar graph) *A graph $G = (N, E)$ is planar if it can be drawn in the plane in such a way that no two edges in E intersect.*

Note that a graph G can be drawn in several different ways; a graph is planar if there exists at least one way of drawing it in the plane in such a way that no two edges cross each other (see Figure A.2).

Definition A.1.15 (Cubic graph) *A graph $G = (N, E)$ is cubic if all its nodes have degree 3.*

Definition A.1.16 (Connected and strongly connected graph) *A graph $G = (N, E)$ is connected if for any two nodes u, $v \in E$ there exists a path from u to v in G. If G is directed, we say that G is strongly connected if for any two nodes u, $v \in E$ there exist a path from u to v, and a path from v to u in G.*

Definition A.1.17 (k-connected and k-edge-connected graph) *A graph $G = (N, E)$ is k-(node-)connected, for some $k \geq 2$, if removing any $k - 1$ nodes from the graph does not disconnect it. Similarly, G is k-edge-connected, for some $k \geq 2$, if removing any $k - 1$ edges from the graph does not disconnect it.*

It can be easily proven that a graph is k-connected if and only if there exist at least k node-disjoint paths between any pair of distinct nodes in G. Similarly, a graph is k-edge-connected if and only if there exist at least k edge-disjoint paths between any pair of distinct nodes in G.

Definition A.1.18 (Graph connectivity and edge connectivity) *The (node) connectivity of a graph $G = (N, E)$, denoted as $\kappa(G)$, is the maximum value of k such that G is k-connected. Similarly, the edge connectivity of G, denoted as $\lambda(G)$, is the maximum value of k such that G is k-edge-connected.*

Theorem A.1.19 *Given a graph $G = (N, E)$, and denoting by $deg_{min}(G)$ the minimal degree of the nodes in N, we have:*

$$\kappa(G) \leq \lambda(G) \leq deg_{min}(G).$$

Definition A.1.20 (Weighted graph) *A weighted graph is a graph in which edges, or nodes, or both, are labeled with a weight.*

Definition A.1.21 (Minimum-cost biconnectivity) *A weighted graph $G = (N, E)$ is minimum-cost biconnected if and only if for any node pair $u, v \in N$ there exists a path connecting u and v in the subgraph G' of G obtained by removing all the nodes in $MP - \{u, v\}$, where MP is the path of minimum cost connecting u and v in G.*

Definition A.1.22 (Monotone graph property) *A certain property $\mathcal{P}$ of a graph is said to be monotone if the fact that $\mathcal{P}$ is satisfied in G implies that $\mathcal{P}$ is satisfied in any supergraph G' of G obtained by adding some edges to G.*

An example of monotone graph property is connectivity: if a certain graph G is connected, then any graph G' obtained from G by adding some edges is also connected.

Definition A.1.23 (Dominating set) *Given a graph $G = (N, E)$, a dominating set for G is a set D of nodes such that for any $u \in N - D$ there exists $v \in D$ such that $(u, v) \in E$, that is, any node in the graph is either in D or adjacent to at least one node in D.*

Definition A.1.24 (Connected dominating set) *Given a graph $G = (N, E)$ and a dominating set D for G, D is said to be a connected dominating set if G_D is connected, that is, if the subgraph of G induced by node set D is connected.*

The examples reported in Figure A.3 clarify the notion of dominating set and connected dominating set.

Definition A.1.25 (Tree) *A tree $T = (N, E)$ is a connected graph with n nodes and $n - 1$ edges, that is, a tree is a minimally connected graph.*

Definition A.1.26 (Rooted tree) *A rooted tree $T = (N, E)$ is a tree in which one of the nodes is selected as the tree root. Once the root node r is chosen, the other nodes in the tree can be classified as either internal node or leaf node. An internal node u is such that there exists $v \in N$ such that $(u, v) \in E$ and $dist(u, r) < dist(v, r)$. A leaf node l is such that, for any $v \in N$ such that $(l, v) \in E$, we have $dist(l, r) > dist(v, r)$.*

Definition A.1.27 (Spanning tree) *Given a connected graph $G = (N, E)$, a spanning tree of G is a tree $T = (N, E_T)$ that contains all the nodes in G and is such that $E_T \subseteq E$.*

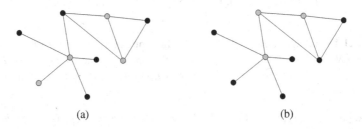

(a)   (b)

Figure A.3 Examples of dominating set (a) and connected dominating set (b). The nodes in the dominating set are represented in light gray.

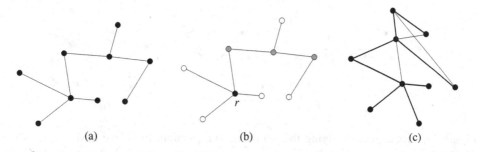

Figure A.4 Examples of tree (a), rooted tree (b), and spanning tree (c). In the rooted tree, internal nodes are gray and leaf nodes are white. The spanning tree on the right is formed by the bold edges.

Figure A.4 reports examples of a tree, a rooted tree, and a spanning tree.

Definition A.1.28 (Cost of a spanning tree) *Given an edge-weighted graph $G = (N, E)$, the cost of a spanning tree T of G is the sum of the weights on its edges.*

Definition A.1.29 (Minimum spanning tree) *Given an edge-weighted graph $G = (N, E)$, a Minimum Spanning Tree (MST) for G is a spanning tree of G of minimum cost.*

Definition A.1.30 (Euclidean MST) *Given a set N of nodes placed in the d-dimensional space (with $d = 1, 2, 3$), and a set of edges E between these nodes, a Euclidean MST (EMST) is a MST of the edge-weighted graph $G = (N, E)$, where each edge has a weight equal to the Euclidean distance between its endpoints.*

Definition A.1.31 (Communication graph) *Given a set N of nodes (representing units of an ad hoc or sensor network), the communication graph is the directed graph $G = (N, E)$ such that edge $(u, v) \in E$ only if v is within u's transmitting range at the current transmit power level.*

Definition A.1.32 (Maxpower graph) *Given a set N of nodes (representing units of an ad hoc or sensor network), the maxpower graph is the communication graph $G = (N, E)$ such that $(u, v) \in E$ if and only if v is within u's transmitting range at maximum power, that is, the maxpower graph contains all possible wireless links between the nodes in the network.*

A.2 Proximity Graphs

Proximity graphs are a class of graphs introduced in the theory of Computational Geometry that are based on proximity relationships between nodes.

Definition A.2.1 (K-neighbors graph) *Given a set N of points in the d-dimensional space, with $d = 1, 2, 3$, and an integer $k \geq 1$, the k-neighbors graph is the directed graph $G_k = (N, E_k)$, where $(u, v) \in E_k$ if and only if v is one of the k closest neighbors of node u.*

Definition A.2.2 (Maximal planar subdivision) *Given a set N of points in the plane, a maximal planar subdivision of N is a planar graph $G = (N, E)$ such that no edge connecting two nodes in N can be added to E without compromising graph planarity.*

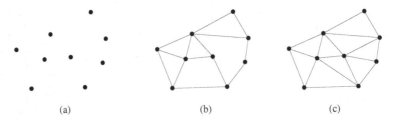

Figure A.5 Examples clarifying the notion of triangulation. In (a) we have a node set N. The graph (b) is a planar subdivision of N, but it is not maximal: in fact, more edges can be added to the graph without compromising planarity. The graph (c) is a triangulation of N.

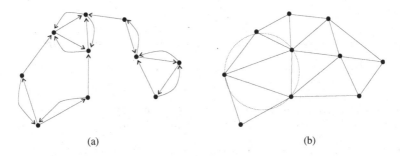

Figure A.6 K-neighbors graph of parameter $k = 2$ (a), and Delaunay triangulation (b). In the Delaunay triangulation, the circumcircle of every triangle (dashed circle) contains no nodes in its interior.

Definition A.2.3 (Triangulation) *Given a set N of points in the plane, a triangulation of N is a maximal planar subdivision whose node set is N.*

Figure A.5 clarifies the notion of triangulation of a set of points.

Definition A.2.4 (Delaunay triangulation) *Given a set N of points in the plane, the Delaunay triangulation of N is the unique triangulation DT of N such that the circumcircle of every triangle contains no points of N in its interior.*

The k-neighbors graph and Delaunay triangulation of a set of points in the plane are reported in Figure A.6.

Definition A.2.5 (Relative neighborhood graph) *Given a set N of points in the plane, the Relative Neighborhood Graph (RNG) of N is the graph $RNG = (N, E)$ such that $(u, v) \in E$ if and only if $lune(u, v)$ does not contain any other point of N in its interior, where $lune(u, v)$ denotes the moon-shaped region formed as the intersection of the two circles of radius $\delta(u, v)$ centered at u and at v.*

Definition A.2.6 (Gabriel graph) *Given a set N of points in the plane, the Gabriel Graph (GG) of N is the graph $GG = (N, E)$ such that $(u, v) \in E$ if and only if the circle that has segment $\overline{uv}$ as diameter does not contain any other point of N in its interior.*

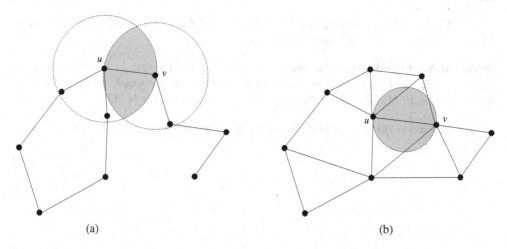

Figure A.7 Relative Neighborhood Graph (a) and Gabriel Graph (b). In the RNG, edge (u, v) exists if and only if $lune(u, v)$ (shaded region) is empty. In the GG, edge (u, v) exists if and only if the circle that has segment $\overline{uv}$ as diameter (shaded region) is empty.

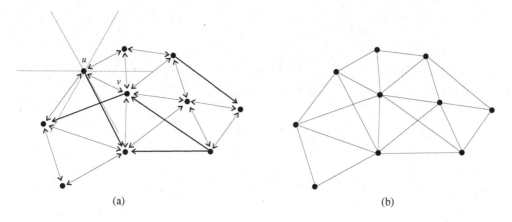

Figure A.8 Yao Graph (a) and Undirected Yao Graph (b). In YG_6, directed edge (u, v) exists if and only if node v is the closest neighbor in one of the cones centered at u. Edges in YG_6 might be unidirectional (bold edges).

The RNG and GG of a set of points in the plane are reported in Figure A.7.

Theorem A.2.7 *Given a set N of points in the plane, we have*

$$EMST \subseteq RNG \subseteq GG \subseteq DT.$$

Definition A.2.8 (Yao graph) *Given a set N of points in the plane, and an integer $k \geq 6$, the Yao Graph of parameter k is the directed graph $YG_k = (N, E_k)$ defined as follows. At each node $u \subset N$, divide the plane into k equally sized cones originating at u. Denoting by*

$C_u^1, \ldots, C_u^k$ the cones for node u, we have that $(u, v) \in E_k$ if and only if there exists cone C_u^i such that v is the closest neighbor of u in C_u^i.

Definition A.2.9 (Undirected Yao graph) *Given a set N of points in the plane, and an integer $k \geq 6$, the Undirected Yao Graph of parameter k is the graph $UYG_k = (N, E_k)$, where $(u, v) \in E_k$ if and only if either edge (u, v) or edge (v, u) is in YG_k.*

The YG and UYG of a set of points in the plane are reported in Figure A.8.

B

Elements of Applied Probability

In this Appendix, we report basic notions of probability theory and briefly describe the main results of some applied probability theories that have been used in the analysis of topology control problems for ad hoc networks. The material of this Appendix is based on (Feller 1957) (Section B.1), on the various papers cited in Section B.2, on (Kolchin et al. 1978) (Section B.3), and on (Meester and Roy 1996) (Section B.4).

B.1 Basic Notions of Probability Theory

Definition B.1.1 (Sample space) *A sample space is the set representing all possible outcomes of a certain random experiment. A sample space is discrete if it is composed of a finite number of elements (e.g. outcomes of a coin toss experiment), or of infinitely many elements that can be arranged into a simple sequence $e_1, e_2, \ldots$.*

Definition B.1.2 (Random variable) *A random variable X is a function defined on a sample space. If the sample space on which X is defined is discrete, X is said to be a discrete random variable.*

Examples of random variables are the number of heads in a sequence of k coin tosses (discrete random variable), the position of a certain particle in a physical system, the position of a sensor thrown from a moving vehicle, and so on.

Definition B.1.3 (Probability distribution) *Let X be a discrete random variable, and let $x_1, x_2, \ldots, x_j, \ldots$ be the possible values of X. The function*

$$P(X = x_i) = f(x_i) \quad (i = 1, 2, \ldots)$$

is called the probability distribution of the random variable X, where $\forall i \ f(x_i) \geq 0$ and $\sum_i f(x_i) = 1$.

Definition B.1.4 (Probability density function) *A probability density function (pdf) on* $\mathbb{R}$ *is a function such that*

$$\forall x \ f(x) \geq 0 \quad \text{and} \quad \int_{-\infty}^{+\infty} f(x)\, dx = 1.$$

Similarly, given a fixed integer $d > 1$, *a pdf on* $\mathbb{R}^d$ *is a function such that*

$$\forall x_1, \ldots, x_d \ f(x_1, \ldots, x_d) \geq 0 \quad \text{and} \quad \int_{-\infty}^{+\infty} f(x_1, \ldots, x_d)\, dx_1 \ldots dx_d = 1.$$

Definition B.1.5 (Continuous random variable) *A random variable* X *taking values in* $\mathbb{R}$ *is continuous if there exists a pdf* f *on* $\mathbb{R}$ *such that*

$$P(a < X \leq b) = \int_a^b f(x)\, dx,$$

for any $a < b$. *Function* f *is called the density of the random variable* X. *A similar definition applies to random variables taking values in* $\mathbb{R}^d$, *for some integer* $d > 1$.

Definition B.1.6 (Distribution function) *Let* $X = (X_1, \ldots, X_d)$ *be a continuous random variable taking values in* $\mathbb{R}^d$, *for some integer* $d \geq 1$. *The function*

$$F(x_1, \ldots, x_d) = P(X_1 \leq x_1, \ldots, X_d \leq x_d) = \int_{-\infty}^{x_1} \ldots \int_{-\infty}^{x_d} f(y_1, \ldots, y_d)\, dy_1 \ldots dy_d,$$

where $f(y_1, \ldots, y_d)$ *is the density of* X, *is called the distribution function of the random variable* X. *Function* f *is called the density of the random variable* X.

Definition B.1.7 (Support of a pdf) *The support of a pdf* f *on* $\mathbb{R}^d$, *for some integer* $d \geq 1$, *is the set of points in* $\mathbb{R}^d$ *on which* f *has positive value. Formally,*

$$supp(f) = \{(x_1, \ldots, x_d) \in \mathbb{R}^d : f(x_1, \ldots, x_d) > 0\}.$$

Clearly,

$$\int_{supp(f)} f(x_1, \ldots, x_d)\, dx_1 \ldots dx_d = 1.$$

Definition B.1.8 (Asymptotic distribution) *A sequence* $X_1, X_2, \ldots, X_n, \ldots$ *of continuous random variables, with distribution functions* $F_1, F_2, \ldots, F_n, \ldots$, *is said to converge in distribution to a certain random variable* X *with distribution* F *if and only if*

$$\lim_{n \to \infty} F_n(x) = F(x)$$

at every continuity point x *of* $F(x)$. *If sequence* $\{X_n\}$ *converges in distribution to a certain random variable* X *with distribution* F, *we say that* F *is the asymptotic distribution of* $\{X_n\}$.

Definition B.1.9 (a.a.s. event) *Let* E_n *be a random variable representing a random event that depends on a certain parameter* n. *We say that the event represented by* E_n *holds asymptotically almost surely (a.a.s.), or with high probability (w.h.p.) if*

$$\lim_{n \to \infty} P(E_n) = 1.$$

Definition B.1.10 (Bernoulli distribution) *A Bernoulli random variable X_p of parameter p, with $0 \le p \le 1$, is a discrete random variable that has value S (success) with probability p and value F (failure) with probability $1 - p$. The corresponding probability distribution is called Bernoulli distribution of parameter p.*

Definition B.1.11 (Poisson process and distribution) *Let us consider a discrete random variable $X(t)$, counting the number of events (e.g. arrival of telephone calls) occurring in the time interval $[0, t]$. If the following properties hold,*

   1. the probability of occurrence of the observed events does not change with time; and

   2. the probability of occurrence of the observed events does not depend on the number of events occurred so far,

then the correspondent random process is called Poisson process. In a Poisson process, the number of events counted after time t follows the probability function

$$P(X(t) = x) = e^{-\lambda t} \frac{(\lambda t)^x}{x!} \quad \text{for } x = 0, 1, 2, \ldots,$$

for some constant $\lambda > 0$. Parameter λ is called the intensity of the Poisson process. The above probability function is called Poisson distribution of parameter λ. A random variable with Poisson distribution is called Poisson random variable.

Definition B.1.12 (Uniform distribution) *Given an interval $[a, b]$, with $a < b$, the uniform distribution on $[a, b]$ is defined by the following probability density function:*

$$f(x) = \begin{cases} 0 & \text{for } x < a \\ \frac{1}{b-a} & \text{for } a \le x \le b \\ 0 & \text{for } x > b \end{cases}.$$

The uniform distribution on arbitrary d-dimensional rectangles is defined similarly. A random variable with uniform distribution on a certain (d-dimensional) interval is called uniform random variable.

Definition B.1.13 (Normal distribution) *The Normal distribution on $\mathbb{R}$ of mean μ and variance σ^2 is defined by the following probability density function $\mathcal{N}(\mu, \sigma)$:*

$$\mathcal{N}(\mu, \sigma)(x) = \frac{1}{\sigma\sqrt{2\pi}} e^{-(x-\mu)^2/(2\sigma^2)}.$$

The Normal distribution on $\mathbb{R}^d$, for some integer $d > 1$, is defined similarly. A random variable with Normal distribution is called Normal random variable.

Definition B.1.14 (Log-normal distribution) *The Log-normal distribution on $\mathbb{R}$ of parameters μ and σ is defined by the following probability density function:*

$$f(x) = \frac{1}{\sigma\sqrt{2\pi}x} e^{-(\ln x - \mu)^2/(2\sigma^2)}.$$

The Log-normal distribution on $\mathbb{R}^d$, for some integer $d > 1$, is defined similarly. A random variable with Log-normal distribution is called log-normal random variable.

B.2   Geometric Random Graphs

A well-established theory that at first glance seems useful in the analysis of ad hoc/sensor network properties is the theory of *random graphs* (Bollobás 1985; Palmer 1985). In this theory, a graph is formed by inserting a certain number of edges between random nodes in the graph, and several properties of this graph are studied (for instance, connectivity, upper/lower bounds on node degree, and so on).

Unfortunately, random graph theory cannot be directly applied in the investigation of ad hoc/sensor network properties since a fundamental assumption in this model is that the probabilities of edge occurrence in the graph are independent, which is not the case in the context of wireless ad hoc networks. In fact, consider a situation in which three nodes u, v, w are located in such a way that $\delta(u, v) < \delta(u, w)$. With common wireless technologies that use omnidirectional antennas, and disregarding the effect of shadowing and fading on radio signal propagation, if u has a link to w, then it has also a link to the closer node v, implying that the occurrence of edge (u, v) is positively correlated to the occurrence of edge (u, w). Even if we consider more-sophisticated radio channel models that account for shadowing and fading of the radio signal, the fact that node u is able to communicate with node w still has an influence on the likelihood of u having a link to the closer node v.

While traditional random graph theory is not very useful in the theoretical analysis of fundamental ad hoc/sensor network properties, a more recent and still-in-development applied probability theory turns out to be very useful to this purpose: the theory of *Geometric Random Graphs* (GRG).

As the name suggests, the theory of GRG can be seen as an extension to the traditional random graph theory in which the graph is not considered as an abstract entity (set of nodes connected by a number of edges), but as a geometric entity (set of points in the d-dimensional space, connected on the basis of a proximity relation).

In a typical GRG model, a set of n points is distributed according to some pdf in a d-dimensional region R, and asymptotic properties of the resulting node placement for $n \to \infty$ are investigated. Among the properties studied in this theory, we cite the following:

- The *minimum* and *maximum node degree*, in a model in which two nodes are connected in the graph if and only if they are at distance of at most $r(n)$ from each other (note that the connection distance is a function of the number of deployed nodes). See (Appel and Russo 1997a,b; Penrose 1999a).

- The *longest nearest neighbor link*, that is, the value of the longest distance between a node and its closest neighbor. See (Dette and Henze 1989; Penrose 1999b; Steele and Tierney 1986).

- The *length of the shortest path* connecting all the deployed nodes. See (Steele 1981).

- The *total edge length*, the *number of connected components*, and the *critical neighbor number* of the k-neighbors graph, which is obtained by connecting each node to its k closest neighbors. See (Avram and Bertsimas 1993; Penrose and Yukich 2001; Wan and Yi 2004; Xue and Kumar 2004).

- The *total edge length* of the Delaunay triangulation built on the deployed nodes. See (Avram and Bertsimas 1993).

- The *length of the longest edge* of the MST built on the deployed nodes. See (Penrose 1997, 1998, 1999b).

- The *total edge length* of the MST built on the deployed nodes. See (Aldous and Steele 1992; Avram and Bertsimas 1992; Steele 1988; Yukich 2000).

The theory of GRG has been used in several recent papers to study fundamental ad hoc/sensor network properties, such as the critical transmitting range for connectivity and k-connectivity (Bettstetter 2002; Panchapakesan and Manjunath 2001; Santi 2005; Wan and Yi 2004; Yi and Wan 2005; Yi et al. 2003), the critical neighbor number (Wan and Yi 2004; Xue and Kumar 2004), and the cost of the optimal solution to the RA and WSRA problem (Blough et al. 2002).

Before concluding this section, we want to outline two important similarities between the theory of GRG and the traditional random graph theory.

A first similarity is the occurrence of the giant component phenomenon (see Section 4.1 for a description of this phenomenon), which has firstly been observed in traditional random graphs, and recently been proven to also occur in geometric random graphs (provided the nodes are deployed in $\mathbb{R}^d$, with $d \geq 2$).

A second similarity is the expected node degree (i.e. number of neighbors) observed in a.a.s. connected graphs, which is known to be $\Omega(\log n)$ in traditional random graphs. A similar result holds also for GRG, in two different models: (i) the homogeneous model, in which every node is connected to every other node within distance $r(n)$; and (ii) the k-neighbors model, in which every node is connected to its k closest neighbors. In model (i), in case of two-dimensional networks, it has been proven that the minimum value of $r(n)$ that guarantees connectivity w.h.p. is such that the expected number of nodes in $\pi r(n)^2$ (i.e. the expected number of neighbors of a node) is $\Omega(\log n)$. A similar result also holds for one- and three-dimensional networks. In model (ii), it has been proven that $k = \Theta(\log n)$ is a necessary and sufficient condition to ensure connectivity w.h.p. of the k-neighbors graph.

For an exhaustive treatment of the theory of GRG, the reader is referred to (Penrose 2003).

B.3 Occupancy Theory

Another applied probability theory that has been successfully used in the analysis of fundamental ad hoc/sensor network properties is the *occupancy theory* (Kolchin et al. 1978).

In the occupancy theory, it is typically assumed that n balls are thrown independently at random into C urns (or cells), where a ball has the same probability of landing in any cell (equiprobable allocation). Variables C and n are interdependent: commonly, it is assumed that the number C of cells is the independent variable, and n is expressed as a function of C.

Given this setting, the asymptotic distribution of several random variables of interest for $C, n(C) \rightarrow \infty$ has been characterized. Among the studied random variables, we cite

- the *number of empty cells* after all balls have been thrown;

- the *number of trials* before at least k urns are filled with at least one ball;

– the *number of balls* in the *minimally* and in the *maximally occupied* cell after all balls
have been thrown.

Several results have also been extended to the case of nonequiprobable allocation of
balls into cells.

The occupancy theory has been used in the context of wireless ad hoc/sensor networks to
characterize the critical transmitting range for connectivity (Santi and Blough 2002, 2003),
to derive bounds to the network lifetime (Blough and Santi 2002), and to study clustering
algorithm properties (Younis and Fahmy 2004; Younis et al. 2004).

Since the usefulness of occupancy theory in the characterization of ad hoc network
properties is not self-evident, we briefly explain how the asymptotic characterization of the
number of empty cells has been used to provide sufficient conditions for a.a.s. connectivity
of the communication graph.

Assume n nodes are deployed in a certain two-dimensional region R. Suppose R is
subdivided into C equal-sized cells, and assume nodes have the same transmitting range r.
If we further assume that nodes are distributed uniformly at random in R, the problem of
determining the minimum number of nodes to be deployed in order to generate a connected
communication graph w.h.p. can be stated as an occupancy problem, provided the cell size
is appropriately chosen.

Consider the cell lattice reported in Figure B.1. It is simple to see that, with this definition
of the cell side, having at least one node in every cell is a *sufficient* condition for generating
a connected communication graph. Hence, a sufficient condition for connectivity can be

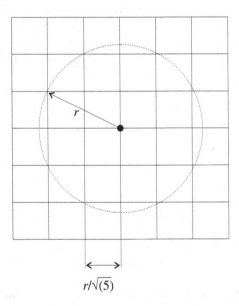

$r/\sqrt{(5)}$

Figure B.1 Cell lattice used to derive a sufficient condition for asymptotic connectivity.
The nodes have transmitting range r, and the cell side is set to $r/\sqrt{5}$. With this definition of
the cell side, having at least one node in each cell is a sufficient condition for connectivity
of the communication graph.

derived by using results from the occupancy theory that determine the asymptotic distribution of number of empty cells. In particular, we have to study the conditions under which this number equals 0, a.a.s. This technique has been used in (Santi and Blough 2002, 2003).

B.4   Continuum Percolation

A third applied probability that proved useful in the analysis of ad hoc/sensor network properties is the theory of *continuum percolation* (Meester and Roy 1996).

In the theory of continuum percolation, nodes are distributed in the plane according to a Poisson process of intensity $\lambda > 0$. In other words, it is assumed that, given any region A of $\mathbb{R}^2$ of area a, the probability of having k nodes in A equals

$$P(A \text{ contains } k \text{ nodes}) = e^{-\lambda a}\frac{(\lambda a)^k}{k!} \quad \text{for } k = 0, 1, 2, \ldots .$$

Once the nodes are deployed, a disk of a certain radius R is centered at each node, and a graph is formed by connecting any two nodes whose disks intersect (see Figure B.2). In the most general model, the radius R_u of the disk centered at node u is a random variable with a certain distribution f (e.g. uniform distribution in $[0, r_{\max}]$), and the random variables R_x modeling the disk radii at the different nodes are assumed to be independent with the same distribution f. In a simpler version of the model, all the disks have the same radius equal to $r > 0$.

Given this setting, researchers have studied the connectivity of the generated graph. In particular, it has been proven that for each $\lambda > 0$ there exists at most one infinite-order connected component in the graph (a.a.s.). However, the existence of an infinite-order component alone is not sufficient to ensure the connectivity of the generated graph.

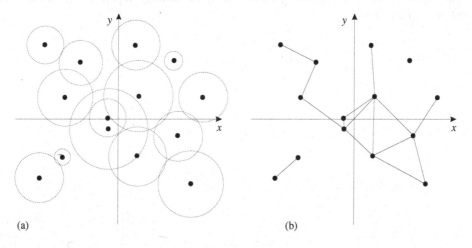

(a)   (b)

Figure B.2   Model used in the continuum percolation theory: nodes are deployed according to a two-dimensional Poisson process of intensity $\lambda > 0$, and a disk is drawn for each node (a); then, a graph is generated by connecting two nodes if and only if their disks intersect (b).

In fact, there could exist (infinitely many) nodes that do not belong to the unbounded component, implying that a significant portion of the nodes are disconnected from the rest of the graph. Hence, the quality of connectivity is related to another parameter, the fraction θ of nodes belonging to the unbounded component, which, in turn, depends on the *percolation probability*. The percolation probability is the probability that an arbitrary node belongs to the connected component of infinite order.

The main result of the theory of continuum percolation is that there exists a finite, positive value λ_c of λ, called the *critical density*, under which the percolation probability is zero, and above which it is nonzero. However, no explicit expression of the percolation probability is known to date, and only bounds to λ_c that hold in some special cases have been derived.

The theory of continuum percolation has been used in the context of wireless ad hoc/ sensor networks to characterize the critical transmitting range for connectivity (Dousse et al. 2003; Dousse et al. 2002; Gupta and Kumar 1998) and to study the message delivery latency of a certain class of sensor networks (Dousse et al. 2004).

The theory of continuum percolation displays not only many similarities with the theory of GRG but also a major difference, that is, the random process used to distribute the nodes: in GRG, a *fixed* number n of nodes is distributed in a certain subregion R of $\mathbb{R}^d$, for some $d \geq 1$, and R is typically *bounded* (e.g. the unit d-dimensional cube); on the contrary, in the theory of continuum percolation, a *random* number of nodes is distributed in an *unbounded* region, which typically is $\mathbb{R}^2$.

In general, deciding which one between the GRG and the continuum percolation models is most suitable to study a certain wireless ad hoc/sensor network property depends on the characteristics of the problem at hand: if we expect to have quite precise information on the number of network nodes and on the size/shape of the deployment region, using the GRG model is preferable; on the contrary, if we have only a vague idea of the number of network nodes and/or of the size/shape of the deployment region, but the expected node density per unit area is approximately known, using the continuum percolation model is the most appropriate choice.

References

Aldous D and Steele J 1992 Asymptotics for euclidean minimal spanning trees on random points. *Probability Theory Related Fields* **92**, 247–258.

Althaus E, Calinescu G, Mandoiu I, Prasad S, Tchervenski N and Zelikovsky A 2003 Power efficient range assignment in ad hoc wireless networks. *Proc. IEEE WCNC 03*, New Orleans, LA, pp. 1889–1894.

Anderegg L and Eidenbenz S 2003 Ad hoc-vcg: a truthful and cost-efficient routing protocol for mobile ad hoc networks with selfish agents. *Proc. ACM Mobicom*, San Diego, CA, pp. 245–259.

Appel M and Russo R 1997a The maximum vertex degree of a graph on uniform points in $[0, 1]^d$. *Advances in Applied Probability* **29**, 567–581.

Appel M and Russo R 1997b The minimum vertex degree of a graph on uniform points in $[0, 1]^d$. *Advances in Applied Probability* **29**, 582–594.

Avram F and Bertsimas D 1992 The minimum spanning tree constant in geometrical probability and under the independent model: a unified approach. *Annals of Applied Probability* **2**(1), 113–130.

Avram F and Bertsimas D 1993 On central limit theorems in geometrical probability. *Annals of Applied Probability* **3**(4), 1003–1046.

Bahramgiri M, Hajiaghayi M and Mirrokni V 2002 Fault-tolerant ad 3-dimensional distributed topology control algorithms in wireless multi-hop networks. *Proc. IEEE International Conference on Computer Communications and Networks*, Dallas, TX, pp. 392–397.

Bai F, Sadagopan N and Helmy A 2003 Important: a framework to systematically analyze the impact of mobility on performance of routing protocols for ad hoc networks. *Proc. IEEE Infocom*, San Francisco, CA, pp. 825–835.

Bao L and Garcia-Luna-Aceves J 2001 Channel access scheduling in ad hoc networks with unidirectional links. *Proc. DIALM 01*, Rome, pp. 9–18.

BenSalem N, Buttyan L, Hubaux J and Jakobsson M 2003 A charging and rewarding scheme for packet forwarding in multi-hop cellular networks. *Proc. ACM MobiHoc*, Annapolis, MD, pp. 13–24.

Bettstetter C 2001a Mobility modeling in wireless networks: categorization, smooth movement, and border effects. *ACM Mobile Computing and Communications Review* **5**(3), 55–67.

Bettstetter C 2001b Smooth is better than sharp: a random mobility model for simulation of wireless networks. *Proc. ACM Workshop on Modeling, Analysis and Simulation of Wireless and Mobile Systems (MSWiM)*, Rome, pp. 19–27.

Bettstetter C 2002 On the minimum node degree and connectivity of a wireless multihop network. *Proc. ACM MobiHoc 02*, Lausanne, pp. 80–91.

Bettstetter C 2004 Failure-resilient ad hoc and sensor networks in a shadow fading environment. *Proc. IEEE Workshop on Dependability Issues in Ad Hoc Networks and Sensor Networks (DIWANS)*, Florence.

Bettstetter C and Krause O 2001 On border effects in modeling and simulation of wireless ad hoc networks. *Proc. IEEE International Conference on Mobile and Wireless Communication Network (MWCN)*, Recife.

Bettstetter C, Resta G and Santi P 2003 The node distribution of the random waypoint mobility model for wireless ad hoc networks. *IEEE Transactions on Mobile Computing* **2**(3), 257–269.

Blough D, Leoncini M, Resta G and Santi P 2002 On the symmetric range assignment problem in wireless ad hoc networks. *Proc. IFIP Conference on Theoretical Computer Science*, Montreal, pp. 71–82.

Blough D, Leoncini M, Resta G and Santi P 2003a *The k-neighbors Approach to Physical Degree Bounded and Symmetric Topology Control in ad hoc networks*. Technical Report IIT-TR-14/2003, Istituto di Informatica e Telematica, Pisa.

Blough D, Leoncini M, Resta G and Santi P 2003b The k-neighbors protocol for symmetric topology control in ad hoc networks. *Proc. ACM MobiHoc 03*, Annapolis, MD, pp. 141–152.

Blough D, Leoncini M, Resta G and Santi P 2003c *K-neighlev: A Practical Realization of Neighborhood-based Topology Control in ad hoc Networks*. Technical Report IIT-TR-09/2003, Istituto di Informatica e Telematica, Pisa.

Blough D, Resta G and Santi P 2004 A statistical analysis of the long-run node spatial distribution in mobile ad hoc networks. *Wireless Networks* **10**, 543–554.

Blough D and Santi P 2002 Investigating upper bounds on network lifetime extension for cell-based energy conservation techniques in stationary ad hoc networks. *Proc. ACM Mobicom*, Atlanta, GA, pp. 183–192.

Bluetooth 1999 Bluetooth specifications available at http://www.bluetooth.org.

Bollobás B 1985 *Random Graphs*. Academic Press, London.

Bollobás B 1998 *Modern Graph Theory*. Springer-Verlag, New York.

Booth L, Bruck J, Cook M and Franceschetti M 2003 Ad hoc wireless networks with noisy links. *Proc. IEEE International Symposium on Information Theory (ISIT)*, Yokohama.

Borbash S and Jennings E 2002 Distributed topology control algorithm for multihop wireless networks. *Proc. IEEE International Joint Conference on Neural Networks*, Honolulu, HI, pp. 355–360.

Bose P, Morin P, Stojmenovic I and Urrutia J 2001 Routing with guaranteed delivery in ad hoc wireless networks. *Wireless Networks* **7**(6), 609–616.

Buchegger S and LeBoudec J 2002a Nodes bearing grudges: towards routing security, fairness and robustness in mobile ad hoc networks. *Proc. Euromicro Workshop on Parallel, Distributed and Network-based Processing*, Canary Islands, pp. 403–410.

Buchegger S and LeBoudec J 2002b Performance analysis of the confidant protocol: cooperation of nodes - fairness in dynamic ad hoc networks. *Proc. ACM Mobihoc*, Lausanne, pp. 226–236.

Burkhart M, Rickenbach PV, Wattenhofer R and Zollinger A 2004 Does topology control reduce interference? *Proc. ACM MobiHoc 04*, Lausanne, pp. 9–19.

Buttyan L and Hubaux J 2001 *Nuglets: A Virtual Currency to Stimulate Cooperation in Self-Organized ad Hoc Networks*. EPFL Technical Report 2001-01-01.

Buttyan L and Hubaux J 2003 Stimulating cooperation in self-organizing mobile ad hoc networks. *Mobile Networks and Applications* **8**(5), 579–592.

Cagali M, Hubaux J and Enz C 2002 Minimum-energy broadcast in all-wireless networks: Np-completeness and distribution issues. *Proc. ACM Mobicom 02*, Atlanta, GA, pp. 172–182.

Calinescu G, Manduiu I and Zelikovsky A 2002 Symmetric connectivity with minimum power consumption in radio networks. *Proc. IFIP Conference on Theoretical Computer Science*, Montreal, pp. 119–130.

Camp T, Boleng J and Davies V 2002 Mobility models for ad hoc network simulations. *Wiley Wireless Communication & Mobile Computing (WCMC)* **2**(5), 483–502.

Cisco 2004 Aironet data sheets available at http://www.cisco.com/en/US/products/hw/wireless.

Clarke E 1971 Multipart pricing of public goods. *Public Choice* **11**, 17–33.

Clementi A, Penna P and Silvestri R 1999 Hardness results for the power range assignment problem in packet radio networks. *Proc. 2nd International Workshop on Approximation Algorithms for Combinatorial Optimization Problems (RANDOM/APPROX'99)*, Berkeley, CA, pp. 197–208.

Couto DD, Aguayo D, Bicket J and Morris R 2003 A high-throughput metric for multi-hop wireless routing. *Proc. ACM Mobicom*, San Diego, CA, pp. 134–146.

CrossBowTechnologies 2004 Motes data sheets available at http://www.xbow.com.

deBerg M, VanKreveld M, Overmars M and Schwarzkopf O 1997 *Computational Geometry: Algorithms and Applications*. Springer-Verlag, Berlin Heidelberg.

Dette H and Henze N 1989 The limit distribution of the largest nearest neighbor link in the unit *d*-cube. *Journal of Applied Probability* **26**, 67–80.

Dousse O, Baccelli F and Thiran P 2003 Impact of interferences on connectivity in ad hoc networks. *Proc. IEEE Infocom 03*, San Francisco, CA, pp. 1724–1733.

Dousse O, Mannersalo P and Thiran P 2004 Latency of wireless sensor networks with uncoordinated power saving mechanisms. *Proc. ACM Mobihoc*, Tokyo, pp. 109–120.

Dousse O, Thiran P and Hasler M 2002 Connectivity in ad hoc and hybrid networks. *Proc. IEEE Infocom 02*, New York, pp. 1079–1088.

Ebert J, Aier S, Kofahl G, Becker A, Burns B and Wolisz A 2002 *Measurement and Simulation of the Energy Consumption of a Wlan Interface*. Technical Report TKN-02-010, Technical University of Berlin.

Efstathiou E and Polyzos G 2005 Fully self-organized fair peering of wireless hotspots. *Wiley European Transactions on Telecommunications*, to appear.

Eidenbenz S, Kumar V and Zust S 2003 Equilibria in topology control games for ad hoc networks. *Proc. ACM DIALM-POMC*, San Diego, CA, pp. 2–11.

Eidenbenz S, Resta G and Santi P 2005 Commit: a sender-centric, truthful, and energy-efficient routing protocol for ad hoc networks. *Proc. IEEE Workshop on Algorithms for Wireless, Mobile, Ad Hoc and Sensor Networks (WMAN)*, Denver, CO to appear.

Faragó A 2002 Scalable analysis and design of ad hoc networks via random graph theory. *Proc. ACM DIAL-M 02*, Atlanta, GA, pp. 43–50.

Feeney L and Nilsson M 2001 Investigating the energy efficiency of a wireless network interface in an ad hoc networking environment. *Proc. IEEE Infocom*, Anchorage, AK, pp. 1548–1557.

Feigenbaum J, Papadimitriou C, Sami R and Shenker S 2002 A bgp-based mechanism for lowest-cost routing. *Proc. ACM Conference on Principles of Distributed Computing (PODC)*, Monterey, CA, pp. 173–182.

Feller W 1957 *An Introduction to Probability Theory and its Applications*. John Wiley and Sons, New York.

Franceschetti M, Booth L, Cook M, Meester R and Bruck J 2005 Continuum percolation with unreliable and spread-out connections. *Journal of Statistical Physics* **118**(3/4).

Frost and Sullivan 2003 Wireless sensors and integrated wireless sensor networks. *Frost and Sullivan Report*.

Gao J, Guibas L, Hershberger J, Zhang L and Zhu A 2001 Geometric spanners for routing in mobile networks. *Proc. ACM MobiHoc 01*, Long Beach, CA, pp. 45–55.

Garcia-Luna-Aceves J and Behrens J 1995 Distributed scalable routing based on vectors of link states. *IEEE Journal Selected Areas in Communication* **13**(8), 1383–1395.

Garey M and Johnson D 1977 The rectilinear steiner tree problem is np-complete. *SIAM Journal of Applied Mathematics* **32**, 826–834.

Girod L and Estrin D 2001 Robust range estimation using acoustic and multimodal sensing. *Proc. IEEE/RSJ International Conference on Intelligent Robots and Systems*, Maui, HI.

Goodman J and O'Rourke J 1997 *Handbook of Discrete and Computational Geometry*. CRC Press, New York.

Grossglauser M and Tse D 2001 Mobility increases the capacity of ad hoc wireless networks. *Proc. IEEE Infocom 01*, Anchorage, AK, pp. 1360–1369.

Groves T 1973 Incentive in teams. *Econometrica* **41**, 617–663.

Gupta P and Kumar P 1998 Critical power for asymptotic connectivity in wireless networks. *Stochastic Analysis, Control, Optimization and Applications*, Birkhauser, Boston, MA, pp. 547–566.

Gupta P and Kumar P 2000 The capacity of wireless networks. *IEEE Transactions Information Theory* **46**(2), 388–404.

Haas Z and Pearlman M 1998 The performance of query control schemes for the zone routing protocol. *Proc. ACM Sigcomm*, Vancouver, pp. 167–177.

Hajek B 1983 Adaptive transmission strategies and routing in mobile radio networks. *Proc. Conference on Information Sciences and Systems*, Baltimore, MD, pp. 373–378.

Hajiaghayi M, Immorlica N and Mirrokni V 2003 Power optimization in fault-tolerant topology control algorithms for wireless multi-hop networks. *Proc. ACM Mobicom 03*, San Diego, CA, pp. 300–312.

Holst L 1980 On multiple covering of a circle with random arcs. *Journal of Applied Probability* **16**, 284–290.

Hong X, Gerla M, Pei G and Chiang C 1999 A group mobility model for ad hoc wireless networks. *Proc. ACM MSWiM*, Seattle, WA, pp. 53–60.

Hou T and Li V 1986 Transmission range control in multihop packet radio networks. *IEEE Transactions on Communications* **COM-34**, 38–44.

Huang Z, Shen C, Srisathapornphat C and Jaikaeo C 2002 Topology control for ad hoc networks with directional antennas. *Proc. IEEE International Conference on Computer Communications and Networks*, Dallas, TX, pp. 16–21.

IEEE 1999 Ansi/IEEE Standard 802.11: Medium Access Control and Physical Specifications. Sect. 15.

IEEE 2004 Ieee antennas & propagation society http://www.ieeeaps.org.

Jain K, Padhye J, Padmanabhan V and Qiu L 2003 Impact of interference on multi-hop wireless network performance. *Proc. ACM Mobicom*, San Diego, CA, pp. 66–80.

Jardosh A, Belding-Royer E, Almeroth K and Suri S 2003 Towards realistic mobility models for mobile ad hoc networks. *Proc. ACM Mobicom*, pp. 217–229.

Johnson D and Maltz D 1996 Dynamic source routing in ad hoc wireless networks. *Mobile Computing*. Kluwer Academic Publishers, pp. 153–181.

Johnson D, Maltz D, Hu Y and Jetcheva J 2002 *The Dynamic Source Routing Protocol for Mobile ad hoc Networks (dsr)*. Internet Draft draft-ietf-manet-dsr-7.txt, Mobile Ad Hoc Networking Working Group.

Jung A and Vaidya N 2002 A power control mac protocol for ad hoc networks. *Proc. ACM Mobicom*, Atlanta, GA, pp. 36–47.

Karp B and Kung H 2000 Gpsr: greedy perimeter stateless routing for wireless networks. *Proc. ACM Mobicom*, Boston, MA, pp. 243–254.

Kawadia V and Kumar P 2003 Power control and clustering in ad hoc networks. *Proc. IEEE Infocom*, San Francisco, CA, pp. 459–469.

Khun F, Wattenhofer R and Zollinger A 2003 Worst-case optimal and average-case efficient geometric ad hoc routing. *Proc. ACM MobiHoc*, Annapolis, MD, pp. 267–278.

Kim D, Toh C and Choi Y 2001 On supporting link asymmetry in mobile ad hoc networks. *Proc. IEEE Globecom 01*, San Antonio, TX, pp. 2798–2803.

Kirousis L, Kranakis E, Krizanc D and Pelc A 2000 Power consumption in packet radio networks. *Theoretical Computer Science* **243**, 289–305.

Kleinrock L and Silvester JA 1978 Optimum transmission radii for packet radio networks or why six is a magic number. *Proc. IEEE National Telecommunication Conference*, Birmingham, AL, pp. 431–435.

Kolchin V, Sevast'yanov B and Chistyakov V 1978 *Random Allocations*. V.H. Winston and Sons, Washington, DC.

Krizman K, Biedka T and Rappaport T 1997 Wireless position location: fundamentals, implementation strategies, and source of error. *Proc. IEEE Vehicular Technology Conference (VTC)*, Phoenix, AZ, pp. 919–923.

Li L, Halpern J, Bahl P, Wang Y and Wattenhofer R 2001 Analysis of a cone-based distributed topology control algorithm for wireless multi-hop networks. *Proc. ACM PODC 01*, Newport, RI, pp. 264–273.

Li N and Hou J 2004 Flss: a fault-tolerant topology control algorithm for wireless networks. *Proc. ACM Mobicom 04*, Philadelphia, PA, pp. 275–286.

Li N, Hou J and Sha L 2003 Design and analysis of an mst-based topology control algorithm. *Proc. IEEE Infocom 03*, San Francisco, CA, pp. 1702–1712.

Li X, Wan P, Wang Y and Frieder O 2002 Sparse power efficient topology for wireless networks. *Proc. IEEE Hawaii International Conference on System Sciences (HICSS)*, Big Island, HI.

Li X, Wang Y and Song W 2004 Applications of k-local mst for topology control and broadcasting in wireless ad hoc networks. *IEEE Transactions on Parallel and Distributed Systems* **15**(12), 1057–1069.

Liang W 2002 Constructing minimum-energy broadcast trees in wireless ad hoc networks. *Proc. ACM Mobihoc 02*, Lausanne. pp. 112–122.

Liu J and Li B 2002 Mobilegrid: capacity-aware topology control in mobile ad hoc networks. *Proc. IEEE International Conference on Computer Communications and Networks*, Dallas, TX, pp. 570–574.

Lynch N 1996 *Distributed Algorithms*. Morgan Kaufmann, San Mateo, CA.

Mainwaring A, Polastre J, Szewczyk R, Culler D and Anderson J 2002 Wireless sensor networks for habitat monitoring. *Proc. ACM Workshop on Wireless Sensor Networks and Applications (WSNA)*, Atlanta, GA, pp. 88–97.

Marina M and Das S 2002 Routing performance in the presence of unidirectional links in multihop wireless networks. *Proc. ACM Mobihoc 02*, Lausanne, pp. 12–23.

Marti S, Giuli T, Lai K and Baker M 2000 Mitigating routing misbehavior in mobile ad hoc networks. *Proc. ACM Mobicom 00*, Boston, MA, pp. 255–265.

Mas-Colell A, Whinston M and Green J 1995 *Microeconomic Theory*. Oxford University Press, New York.

Meester R and Roy R 1996 *Continuum Percolation*. Cambridge University Press, Cambridge, MA.

MeyerAufDerHeide F, Schindelhauer C and Grunewald M 2002 Congestion, dilation, and energy in radio networks. *Proc. ACM Symposium on Parallel Algorithms and Architectures (SPAA)*, Winnipeg, pp. 10–13.

Michiardi P and Molva R 2002 Core: a collaborative repudiation mechanism to enforce node cooperation in mobile ad hoc networks. *Proc. IFIP Conference on Security, Communications and Multimedia*, Washington, DC.

Moaveni-Nejad K and Li X 2005 Low-interference topology control for wireless ad hoc networks. *Ad Hoc and Sensor Networks: an International Journal*, **1**(1–2), pp. 41–64.

Narayanaswamy S, Kawadia V, Sreenivas R and Kumar P 2002 Power control in ad hoc networks: theory, architecture, algorithm and implementation of the compow protocol. *Proc. European Wireless 2002*, Florence, pp. 156–162.

Ni S, Tseng Y, Chen Y and Sheu J 1999 The broadcast storm problem in mobile ad hoc networks. *Proc. ACM Mobicom*, Seattle, WA, pp. 151–162.

Niculescu D and Nath B 2003 Ad hoc positioning system (aps) using aoa. *Proc. IEEE Infocom 03*, San Francisco, CA, pp. 1734–1743.

Nisan N and Ronen A 2001 Algorithmic mechanism design. *Games and Economic Behavior* **35**, 166–196.

Ns2 2002 The network simulator - ns-2 http://www.isi.edu/nsnam/ns/.

Osborne M and Rubinstein A 1994 *A Course in Game Theory*. MIT Press, Binghamton, NY.

Palmer E 1985 *Graphical Evolution*. John Wiley and Sons, New York.

Panchapakesan P and Manjunath D 2001 On the transmission range in dense ad hoc radio networks. *Proc. IEEE SPCOM 2001*, Bangalore.

Pearlman M, Haas Z and Manvell B 2000a Using multi-hop acknowledgements to discover and reliably communicate over unidirectional links in ad hoc networks. *Proc. Wireless Communications and Networking Conference (WCNC)*, Chicago, IL, pp. 532–537.

Pearlman M, Haas Z, Sholander P and Tabrizi S 2000b On the impact of alternate path routing for load balancing in mobile ad hoc networks. *Proc. ACM MobiHoc*, Boston, MA, pp. 3–10.

Penrose M 1997 The longest edge of the random minimal spanning tree. *The Annals of Applied Probability* **7**(2), 340–361.

Penrose M 1998 Extremes for the minimal spanning tree on normally distributed points. *Advances in Applied Probability* **30**, 628–639.

Penrose M 1999a On k-connectivity for a geometric random graph. *Random Structures and Algorithms* **15**(2), 145–164.

Penrose M 1999b A strong law for the largest nearest-neighbour link between random points. *Journal of London Mathematical Society* **60**(2), 951–960.

Penrose M 1999c A strong law for the longest edge of the minimal spanning tree. *The Annals of Probability* **27**(1), 246–260.

Penrose M 2003 *Random Geometric Graphs*. Oxford University Press, Oxford.

Penrose M and Yukich J 2001 Central limit theorems for some graphs in computational geometry. *Annals of Applied Probability* **11**, 1005–1041.

Perkins C, Belding-Royer E and Das S 2002 Ad hoc on-demand distance vector (aodv) routing. Internet Draft Draft-ietf-manet-aodv-12.txt, Mobile Ad Hoc Networking Working Group.

Perkins C and Bhagwat P 1994 Highly dynamic destination-sequenced distance vector routing (dsdv) for mobile computers. *Proc. ACM SigComm*, London.

Philips T, Panwar S and Tantawi A 1989 Connectivity properties of a packet radio network model. *IEEE Transactions Information Theory* **35**(5), 1044–1047.

Piret P 1991 On the connectivity of radio networks. *IEEE Transactions Information Theory* **37**(5), 1490–1492.

Pister K 2001 The smart dust project http://robotics.eecs.berkeley.edu/~pister/SmartDust/.

Polastre J, Szewczyk R, Sharp C and Culler D 2004 The mote revolution: low power wireless sensor network devices. *Proc. Hot Chips 16: A Symposium on High Performance Chips*, Stanford, CA.

Prakash R 2001 A routing algorithm for wireless ad hoc networks with unidirectional links. *ACM/Kluwer Wireless Networks* **7**(6), 617–625.

Prim R 1957 Shortest connection networks and some generalizations. *The Bell System Technical Journal* **36**, 1389–1401.

Priyantha N, Chakraborty A and Balakrishnan H 2000 The cricket location-support system. *Proc. ACM Mobicom 00*, Boston, MA, pp. 32–43.

Raghunathan V, Schurgers C, Park S and Srivastava M 2002 Energy-aware wireless microsensor networks. *IEEE Signal Processing Magazine* **19**(2), 40–50.

Ramanathan R and Rosales-Hain R 2000 Topology control of multihop wireless networks using transmit power adjustment. *Proc. IEEE Infocom 00*, Tel Aviv, pp. 404–413.

Ramanathan S and Steenstrup M 1998 Hierarchically-organized multi-hop mobile networks for multimedia support. *Mobile Networks and Applications* **3**(1), 101–119.

Ramasubramanian V, Chandra R and Mosse D 2002 Providing a bidirectional abstraction for unidirectional ad hoc networks. *Proc. IEEE Infocom 02*, New York, pp. 1258–1267.

Rappaport T 2002 *Wireless Communications: Principles and Practice*, Second Edition. Prentice Hall, Upper Saddle River, NJ.

RockwellScienceCenter 2004 The wins project available at http://wins.rsc.rockwell.com.

Rodoplu V and Meng T 1999 Minimum energy mobile wireless networks. *IEEE Journal Selected Areas in Communication* **17**(8), 1333–1344.

Royer E, Melliar-Smith P and Moser L 2001 An analysis of the optimum node density for ad hoc mobile networks. *Proc. IEEE International Conference on Communications*, Helsinki, pp. 857–861.

Santi P 2005 The critical transmitting range for connectivity in mobile ad hoc networks. *IEEE Transactions on Mobile Computing*, **4**(3), pp. 310–317.

Santi P and Blough D 2002 An evaluation of connectivity in mobile wireless ad hoc networks. *Proc. IEEE DSN 2002*, Bethesda, MD, pp. 89–98.

Santi P and Blough D 2003 The critical transmitting range for connectivity in sparse wireless ad hoc networks. *IEEE Transactions on Mobile Computing* **2**(1), 25–39.

Savvides A, Han C and Srivastava M 2001 Dynamic fine-grained localization in ad hoc networks of sensors. *Proc. ACM Mobicom 01*, Rome, pp. 166–179.

Sen A and Huson M 1996 A new model for scheduling packet radio networks. *Proc. IEEE Infocom*, San Francisco, CA, pp. 1116–1124.

Shih E, Bahl P and Sinclair M 2002 Wake on wireless: an event driven energy saving strategy for battery operated devices. *Proc. ACM Mobicom*, Atlanta, GA, pp. 160–171.

Song W, Wang Y, Li X and Frieder O 2004 Localized algorithms for energy efficient topology in wireless ad hoc networks. *Proc. ACM MobiHoc*, Tokyo, pp. 98–108.

Steele J 1981 Subadditive euclidean functionals and nonlinear growth in geometric probability. *Annals of Probability* **9**(3), 365–376.

Steele J 1988 Growth rates of euclidean minimal spanning trees with power weighted edges. *Annals of Probability* **16**, 1767–1787.

Steele J and Tierney L 1986 Boundary domination and the distribution of the largest nearest neighbor link in higher dimensions. *Journal of Applied Probability* **23**, 524–528.

Takagi H and Kleinrock L 1984 Optimal transmission ranges for randomly distributed packet radio terminals. *IEEE Transactions on Communications* **COM-32**, 246–257.

Vickrey W 1961 Counterspeculation, auctions, and competitive sealed tenders. *Journal of Finance* **16**, 8–37.

Wan P, Calinescu G, Li X and Frieder O 2002 Minimum energy broadcasting in static ad hoc wireless networks. *ACM/Kluwer Wireless Networks* **8**(6), 607–617.

Wan P and Yi C 2004 Asymptotical critical transmission radius and critical neighbor number for k-connectivity in wireless ad hoc networks. *Proc. ACM MobiHoc 04*, Tokyo, pp. 1–8.

Wang K and Li B 2002 Group mobility and partition prediction in wireless ad hoc networks. *Proc. IEEE International Conference on Communications*, New York, pp. 1017–1021.

Wang W and Li X 2004 Truthful multicast routing in selfish wireless networks *Proc. ACM Mobicom*, Philadelphia, PA, pp. 245–259.

Wang Y, Li X and Frieder O 2002 Distributed spanners with bounded degree for wireless ad hoc networks. *International Journal of Foundations of Computer Science*, **14**(2), pp. 183–200.

Wang X, Xing G, Zhang Y, Lu C, Pless R and Gill C 2003 Integrated coverage and connectivity configuration in wireless sensor networks. *Proc. ACM SenSys*, Los Angeles, CA, pp. 28–39.

Wattenhofer R, Li L, Bahl P and Wang Y 2001 Distributed topology control for power efficient operation in multihop wireless ad hoc networks. *Proc. IEEE Infocom 01*, Anchorage, Alaska, pp. 1388–1397.

Wattenhofer R and Zollinger A 2004 Xtc: A practical topology control algorithm for ad hoc networks. *4th International Workshop on Algorithms for Wireless, Mobile, Ad Hoc and Sensor Networks (WMAN)*. Santa Fe, NM.

Wieselthier J, Nguyen G and Ephremides A 2000 On the construction of energy-efficient broadcast and multicast trees in wireless networks. *Proc. IEEE Infocom 00*, Tel Aviv, pp. 585–594.

Xue F and Kumar P 2004 The number of neighbors needed for connectivity of wireless networks. *Wireless Networks* **10**(2), 169–181.

Yi C and Wan P 2005 Asymptotic critical transmission ranges for connectivity in wireless ad hoc networks with bernoulli nodes. *Proc. IEEE Wireless Communications and Networking Conference (WCNC)*, New Orleans, LA (to appear).

Yi C, Wan P, Li X and Frieder O 2003 Asymptotic distribution of the number of isolated nodes in wireless ad hoc networks with bernoulli nodes. *Proc. IEEE Wireless Communications and Networking Conference (WCNC)*, New Orleans, LA, pp. 1585–1590.

Yoon J, Liu M and Noble B 2003 Random waypoint considered harmful. *Proc. IEEE Infocom*, San Francisco, CA, pp. 1312–1321.

Younis O and Fahmy S 2004 Heed: a hybrid, energy-efficient, distributed clustering approach for ad hoc sensor networks. *IEEE Transactions on Mobile Computing* **3**(4), 366–379.

Younis O, Fahmy S and Santi P 2004 Robust communications for sensor networks in hostile environments. *Proc. IEEE Internation Workshop on Quality of Service (IWQoS)*, Montreal, pp. 10–19.

Yukich J 2000 Asymptotics for weighted minimal spanning trees on random points. *Stochastic Processes and their Application* **85**, 123–128.

Zeng X, Bagrodia R and Gerla M 1998 Glomosim: a library for parallel simulation of large-scale wireless networks. *Workshop on Parallel and Distributed Simulations (PADS)*, Banff, Alberta.

Zhong S, Yang Y and Chen J 2003 Sprite: a simple, cheat-proof, credit-based system for mobile ad hoc networks. *Proc. IEEE Infocom*, San Francisco, CA, pp. 1987–1997.

ZigBeeAlliance 2004 http://www.zigbee.org.

Index

Printed in the United States
By Bookmasters